AF443051

CORE CONCEPT OF
NANOTECHNOLOGY
WITH APPLICATION SPECTRUM

CORE CONCEPT OF
NANOTECHNOLOGY
WITH APPLICATION SPECTRUM

Edited by
Rakesh Rathi

2007

SBS Publishers & Distributors Pvt. Ltd.
New Delhi

All rights reserved. No part of this publication may be reproduced, stored in a retrieval system, or transmitted in any form or by any means, electronic, mechanical, photocopying, recording or otherwise, without the prior written permission of the publisher and the copyright holder.

ISBN10: 81-89741-50-0
ISBN13: 978-81-89741-50-1

Indian Price - Rs 895
Foreign Price - USD 50

First Published in India in 2007

© Reserved

Published by:
SBS PUBLISHERS & DISTRIBUTORS PVT. LTD.
2/9, Ground Floor, Ansari Road, Darya Ganj,
New Delhi - 110002, INDIA
Tel: 23289119, 41563911
Email: mail@sbspublishers.com

Printed in India by SALASAR IMAGING SYSTEMS, New Delhi.

PREFACE

Our future lies in Nanotechnology.

The next wave of global advances in IT and Telecom, medicine, security will be driven by innovations in nanotechnology.

If you want to be a part of tiniest miracle you must understand core concepts and applications of nanotechnology.

This book attempts to help you to achieve basic understanding of the subject.

The book covers 95 per cent of the subject matter taught first in the vast majority of Universities.

The understanding of any subject is inevitably dependent on the person stydying, howerver, access to study material that has simplicity on explanation and directions to furtehr information is invaluable; this book has tried to incorporate thest two requirements.

This book does not cover detail technical analysis of nanotechnology tools.

Many thanks to those whos have helped me in editing this book. Thanks to Mr. Bhasin for encouragement whilst I was editing the book.

Er. Rakesh Rathi

CONTENTS

1

INTRODUCTION

Manufactured products are made from atoms. The properties of those products depend on how those atoms are arranged. If we rearrange the atoms in coal we can make diamond. If we rearrange the atoms in sand (and add a few other trace elements) we can make computer chips. If we rearrange the atoms in dirt, water and air we can make potatoes.

Many scientific and technological advances recently made depend on the properties of materials at a *very very* small scale. Such technological advances are known as 'nanotechnology', the prefix 'nano'—being derived from the Greek word 'nanos', which means dwarf.

NANOTECHNOLOGY

The development and practical applications of structures and devices on a nanometer scale (between 1 and 100 nanometers) (A Nanometer is one-billionth of a meter 1 nm = 10^{-9} m)

A dimension of 100 nanometers is important in nanotechnology, because under this limit one observes new properties of matter, primarily due to the laws of quantum physics.

Nanotechnology is currently the subject of much research and study, and as a result it is developing rapidly. It has great potential for producing improvements and innovations in many areas of life, not least of which would be *new and improved health treatments; reduced use of some harmful or scarce resources; cleaner, faster and safer manufacturing; quicker and smaller devices; increased life-cycle of products; many other improvements to existing products.*

Nanotechnology is currently in a very infantile stage. However, we now have the ability to organize matter on the atomic scale and there are already numerous products available as a direct result of our rapidly increasing ability to fabricate and characterize feature sizes less than 100 nm. Mirrors that don't fog, biomimetic paint with a contact angle near 180°, gene chips and fat soluble vitamins in aqueous beverages are some of the first manifestations of nanotechnology. However, it will be the fields of medicine and computer science, where nanotechnology will first show its presence.

Nanoscience is an interdisciplinary field that seeks to bring about mature nanotechnology. Focusing on the nanoscale intersection of fields such as physics, biology, engineering, chemistry, computer science and more, nanoscience is rapidly expanding. Nanotechnology centers are popping up around the world as more funding is provided and nanotechnology market share increases. The rapid progress is apparent by the increasing appearance of the prefix "nano" in scientific journals and the news. Thus, as we increase our ability to fabricate computer chips with smaller features and improve our ability to cure disease at the molecular level, nanotechnology is here.

There's hard and fast evidence that the nanotechnology industry is shifting out of the hype stage into the commercialization stage. According to the National Center for Manufacturing Sciences (NCMS), which studied the adoption of nanotechnology in the U.S. manufacturing industry, 58% of the nearly 600 survey respondents indicate they will have nanomanufactured products on the market within the next three years.

The top commercialized products already on the market include semiconductors, nanowires, coatings, and thin films, according to the NCMS.

Nanotechnology could prove to be a "transformative" technology comparable in its impact to the steam engine in the 18th century, electricity in the 20th century, and the Internet in contemporary society .

Nature is master in operating at the nanoscale-it has always depended on nanoscale operations for the functioning of life, for example: the metabolic activities within cells for sustaining life; the fertilisation of an embryo in reproduction; and the interaction of cellular components for cell movement, to name a few.

The ecosystem is another natural system that is heavily dependent on nanotechnology-nanoparticles are important in the formation of both rain and snow. Small airborne particles known as condensation nuclei act as the nucleation site, or seed site, on which water condenses to form raindrops of snowflakes depending on the temperature. The particles can vary in size from thousands of nanometres down to a few hundreds of nanometers, or less. These nanoparticles can arise from a variety of sources including dust, valcanoes, and forest or factory smoke, phytoplankton or ocean salt.

Much of the photosynthesis that powers forests unfolds inside tiny cellular power houses called chloroplasts (above). These contain nanoscale molecular machinery (including pigment molecules like chlorophyll) arranged inside stacked structures, called thylakoid disks, that convert light and carbon dioxide into biochemical energy.

Even an early instance of nanotechnology like catalysis really is young compared to nature's own nanotechnology, which emerged billions of years ago when molecules began organizing into the complex structures that could support life. *Photosynthesis*, biology's way of harvesting the solar energy that runs so much of the planet's living kingdom, is one of those ancient products of evolution. Scientists often perceive photosynthesis as the result of brilliantly engineered molecular ensemble them which include light-harvesting molecules such as chlorophyll arranged within cells on the nanometer and micrometer scales. These ensembles capture light energy and convert it into the chemical energy (which is stored in chemical bonds) that drives the biochemical machinery of plant cells.

The abalone, a mollusk, serves up another perennial favorite in nature's gallery of enviable nanotechnologies. These squishy creatures construct supertough shells with beautiful, iridescent inner surfaces. They do this by organizing the same calcium carbonate of crumbly schoolroom chalk into tough nanostructured bricks. For mortar, abalones concoct a stretchy goo of protein and carbohydrate. Cracks that may start on the outside rarely make it all the way through; the structure of the shell forces a crack to take a tortuous route around the tiny bricks, which dissipates the energy behind the damage. Adding to the damage control is that stretchy mortar. As a crack grows, the mortar forms resilient nanostrings that try to force any separating bricks back together. The result is a *Lilliputian* masonry that can withstand sharp beaks, teeth, even hammer blows.

This clever engineering of the abalone shell reflects one of nanotechnology's most enticing faces: by creating nanometer-scale structures, it's possible to control the fundamental properties like color, electrical conductivity, melting temperature, hardness, crack-resistance, and strength of materials without changing the materials' chemical composition. The stuff of soft chalk becomes hard shell. Another feat of natural nanotechnology is in continuous operation every time you take a breath, move a muscle, live another second. Known antiseptically as F1-ATPase complexes, they're actually molecular motors inside cells. Each of these motors is a complex of proteins bound to the membranes of mitochondria, the cell's bacteria-sized batteries. About 10 nanometers across, the F1-ATPase complexes are key players in the synthesis of ATP the molecular fuel for cellular activity. Scientists have found that F1-ATPase complexes also generate rotary motion just like fan motors whirring in summertime windows. By attaching tiny protein filaments to the hub of F1-ATPase, researchers have visualized this rotary action. And it sets their own nanoengineering imaginations into motion with designs for human-made nanometer-scale machines.

Scientists using tools like electron microscopes to look at natural structures like abalone shells and protein complexes hope to emulate some of biology's nanoscale engineering. Their aim is to create structural materials for stronger, lighter, more damage-resistant and otherwise better man-made constructions ranging from buildings and cars to batteries and prosthetic limbs.

From chalk to abalone shell... this is the "alchemy" of natural nanotechnology without human intervention. And now physicists, chemists, materials scientists, biologists, mechanical and electrical engineers, and many other specialists are pooling their collective knowledge and tools so that they too can tailor the world on atomic and molecular scales. Abilities like this in the coming century could make the accomplishments of the 20th century seem quaint.

The pyramids in Egypt, the Brooklyn Bridge, and automobiles are conspicuous monuments to how well and how long people have controlled matter on large scales of meters and miles. The products of Swiss watchmakers even several centuries ago proved that human control over the material world had extended downward a thousandfold to the millimeter scale or so. Over the past few decades, researchers have pushed this control down another hundredfold. Using microlithographic techniques, they've learned to inscribe silicon with ultra-dense patterns of circuitry whose individual components now are only visible with powerful electron microscopes.

And all the while, chemists have been learning to mix, blend, heat, react, and otherwise process chemicals to produce millions of different specific molecular structures. This is about the finest level of material structure relevant for making things. In so doing, researchers have developed recipes and protocols for mak-ing the plastics, ceramics, semiconductors, metals, glass, fabrics, composites and other materials of the constructed landscape.

But there's a big gap between the scale of individual molecular structures made by chemists and the the submicroscopic components on microprocessors made by electrical engineers. That gap, which spans from about one nanometer to several hundred nanometers, is where fundamental properties are defined.

So with every advance researchers make in nanotechnology, they stitch together an unbroken engineering nexus from atoms on up to skyscrapers. Roald Hoffmann, a chemist and Nobel Prize laureate at Cornell University has put it this way: "Nanotechnology is the way of ingeniously controlling the building of small and large structures, with intricate properties; it is the way of the future, a way of precise, controlled building, with incidentally, environmental benignness built in by design."

Many researchers expect that better control over the way atoms and molecules assemble into tiny structures could lead to a host of new technologies based on quantum phenomena that become prevalent only at nanometer scales.

In the normal-sized realm of books, bricks, cars and houses, quantum mechanics the conceptual framework scientists use to describe and predict the properties of matter on the levels of atoms and electrons doesn't have much direct relevance. As long as bricks hold up the house, who cares about their quantum specifics? But now researchers actually are creating nanoscale building blocks, such as metallic and ceramic particles, and all-carbon "nanotubes," that are hundreds of millions of times smaller than bricks used for houses and tubes used for plumbing. So scientists are finding themselves in the middle of quantum mechanics territory.

In this nanoscale territory, electrons, for one, no longer flow through electrical conductors like rivers. On this scale, an electron's quantum mechanical nature expresses itself as a wave. This behavior makes it possible for electrons to do remarkable things, such as instantly tunnel through an insulating layer that normally would have stopped it dead. The payoff of this behavior is that electronic devices built on nanoscales not only can pack more densely on a chip but also can operate far faster and with dramatically fewer electrons and less energy loss than conventional transistors.

These characteristics ultimately could yield more powerful computers that can help scientists mimic phenomena better and engineers design products better. In some specially designed materials with nano-scale layers made of different semiconductor materials, electrons exhibit behaviors not possible in less precisely organized settings. Researchers already have exploited this quantum mechanical reality to design and build new solid state lasers that emit light in wavelengths good for tasks like monitoring pollution, tracking chemical reactions, and optical communications.

What's more, researchers already have taken steps toward so-called quantum computers based on the energy states of atoms and electrons. Theorists predict that quantum computers, if they can be realized, will be able to calculate millions, even billions, of times faster than today's super-computers. Nanotechnology may propel early laboratory steps and theoretical projections like these toward practical versions.

Another major fountain of new technology is likely to spout from a simple fact of material reality: as objects become smaller, the proportion of their constituent atoms at or near the surface rises. Collections of very small particles, therefore, have high surface area compared to their volume. This characteristic is profound because so much of what happens in the world happens at surfaces. Photosynthesis occurs on surfaces inside of cells. Catalysis happens on the surfaces of particles. Ice in the atmosphere forms on the surfaces of floating dust specks.

Smaller industrial catalysts, or ones with labyrinthine interiors with nanoscale features, mean there's more surface area for thousands of chemical transformations. Make a ceramic brick or a metal part, such as a jet engine's turbine blade, out of nanoscale powder particles instead of conventional microscale powder particles and the amount of internal surface goes up dramatically. There's no chemical difference between the two materials; only the size of their constituent particles differs. Yet, just for that the nanostructured brick or metal piece may be harder, less likely to crack, or stronger at higher temperatures, than the conventional brick. That's important for people who make things like armor or turbine blades for jet engines, which run more efficiently the hotter they become. The shift to nanoscale building blocks for making metal, ceramic, polymer and other material components is enabling researchers to "dial-in" many properties like melting temperature, magnetic properties (e. g., the magnetic detection ability of materials in the read heads of hard disks), and color that previously were impossible to obtain for a particular material.

Nanotechnology has given us the tools... to play with the ultimate toy box of nature atoms and molecules. Everything is made from it. The possibilities to create new things appear limitless Principal domains which will be affected by developments in Nanotechnology:

- *Materials*: new materials, harder, more durable and resistant, lighter and less expensive.
- *Electronics*: electronic components will become smaller and smaller, allowing the design of more powerful computers.
- *Energy*: a vast increase in the potential of solar energy generation is envisioned, for example.
- *Health and nanobiotechnologies*: great expectations are held in the areas of prevention, diagnostics, and treatment. For example, nanoscopic probes could be put in place to measure our state of health around the clock, new tools could be developed to fight genetic disease at the level of the gene, and markers could be created to detect and, one by one, destroy cancerous cells, just to name a few of the many possibilities.

Developments in these domains would impact a broad range of industries, such as cosmetics, pharmaceuticals, consumer appliances, hygienics, construction, communication, security and safety, and space exploration. Our environment will benefit as well, in terms of clean, economical energy production, and the use of more environmentally friendly materials.

If brief, many areas of our daily lives will be affected in one way or another by the development of Nanotechnologies, because Nanotechnologies will permit us to do better, with less.

As of today, at the dawn of the Third Millennium, nanotechnological products are already on the market. Thus, one can purchase lighter and stronger tennis rackets composed of carbon nanotubes, or even cosmetics containing nanoparticles which allow better penetration of the skin. But we are still far from the nanotechnology era which will impact our daily lives. When will that revolution take place? When will we benefit substantially from progress in Nanotechnology research and development? The estimates vary. The predicted range is from the years 2010 to 2040.

The stakes involved in the development of Nanotechnologies are continental: America, Europe, and Asia are actively preparing for ongoing development efforts which won't be stopping anytime soon. Massive investments are being made for the purpose of developing Nanotechnologies all over the world.

"Nanotechnology is likely to be particularly important in the developing world, because it involves little labour, land or maintenance; it is highly productive and inexpensive; and it requires only modest amounts of materials and energy."

Given the promise of nanotechnology, the race is on to harness its potential—and to profit from it. Many governments believe nanotechnology will bring about a new era of productivity and wealth, and this is reflected by the way public investment in nanotechnology research and development has risen during the past decade. In 2002, Japan was dedicating US$750 million a year to the field, a six-fold increase on the 1997 figure.

The US National Science Foundation predicts that the global market for nanotech-based products will exceed US$1 trillion within 15 years. Paul Miller, senior researcher at the British policy research organisation Demos, said in 2002 that "already, roughly one-third of the research budgets of the biggest science-based firms in the US is going into nanotech" whilst the US National Nanotechnology Initiative's budget rose from US$116 million in 1997 to a requested US$849 million in 2004.

In the developing world, Brazil, Chile, China, India, the Philippines, South Korea, South Africa and Thailand have shown their commitment to nanotechnology by establishing government-funded programmes and research institutes.

Researchers at the University of Toronto Joint Centre for Bioethics have classified these countries as 'front-runners' (China, South Korea, India) and 'middle ground' players (Thailand, Philippines, South Africa, Brazil, Chile). In addition, Argentina and Mexico are 'up and comers': although they have research groups studying nanotechnology, their governments have not yet organised dedicated funding.

In May 2004, the Thai government announced plans to use nanotechnology in one per cent of all consumer products by 2013. Their market value by then is predicted to be 13 trillion baht (more than US$320 billion at contemporary exchange rates). Indeed, Thailand has wholeheartedly embraced nanotechnology and its development is a major commitment of the Thai government. Likewise, China announced in May 2004 that nanotechnology is central to its long-term national science and technology plan.

Nanotechnology holds the promise of new solutions to problems that hinder the development of poor countries, especially in relation to health and sanitation, food security, and the environment.

In its 2005 report entitled Innovation: applying knowledge in development, the UN Millennium Project task force on science technology and innovation wrote that "nanotechnology is likely to be particularly important in the developing world, because it involves little labour, land or maintenance; it is highly productive and inexpensive; and it requires only modest amounts of materials and energy".

Nanotechnology is a classic, general-purpose technology (GPT). Other GPTs, including steam engines, electricity, and railroads, have been the basis for major economic revolutions. GPTs typically start as fairly crude technologies, with limited uses, but then rapidly spread into new applications.

All prior GPTs have led directly to major upheavals in the economy—the process of creative destruction. And nanotechnology may be larger than any of the other GPTs that preceded it. Creative destruction is the process by which a new technology or product provides an entirely new and better solution, resulting in the complete replacement of the original technology or product. Investors should expect that creative destruction will not only continue, but will also likely accelerate, and nanotechnology will be at the core.

What does this mean from a practical standpoint? Because of the advent of nanotechnology, we believe new companies will displace a high percentage of today's leading companies. The majority of the companies in today's Dow Jones industrials Index are unlikely to be there 20 years from now.

Along those same lines, Josh Wolfe of Lux Capital, editor of the Forbes/Wolfe Nanotech Report, writes: "Quite simply, the world is about to be rebuilt (and improved) from the atom up. That means tens of trillions of dollars to be spent on everything: clothing... food...cars...housing...medicine...the devices we use to communicate and recreate...the quality of the air we breathe...and the water we drink, are all about to undergo profound and fundamental change. Nanotechnology will shake up just about every business on the planet."

If the development of Nanotechnologies is to be encouraged, the effort must be made in the right direction: safeguards must be put in place, because as with all great technological advances, new potentialities contain unknowns and risks about which we must be concerned, such as a new arms race based on smaller, deadlier weapons

Major technological revolutions, including the industrial revolution and the dawn of the information era, have revealed how new discoveries can drastically change our lives. There is no doubt that rapid technological transformations require new paradigms of how to educate the next generation of leaders in academia and industry. Definately one method is through good books.

This book will serve as a nano technology handbook for students, teachers, researchers and industrialist.

2

HISTORY

Pottery using nano-sized particles has been in use for thousands of years. The oldest known such object is thought to be the Lycurgus chalice, which dates back to the late fourth century AD, and which can be seen in the British Museum. The Roman chalice has a raised frieze showing the myth of King Lycurgus, and is made from glass which appears green in reflected light, but when light is shone directly through the glass it appears translucent red.

This unusual optical effect is caused by 70nm particles of silver and gold contained within the glass. (The glass contains 40 parts per million of gold and 300 parts per million of silver, and the particles consisting of seven parts silver to three parts gold.) It is possible that, while the glass was being made, some scrap metal and slag containing silver and gold was added, and the unusual effect was achieved by accident. The understanding that glass could be coloured red by adding small amounts of gold is often credited to Johann Kunkel, who worked in Germany in the late seventeenth century.

The chalice, however, is not a unique example of nanoparticles being used in centuries past; pottery from the ninth century AD from the Mesopotamian era also contains nanoparticles in glazed films applied to ceramic pottery. The technique was brought to Spain in medieval times, as Arabian culture spread, and it then migrated to Italy, where Renaissance pottery made much use of the effect in polychrome lustres on pottery.

The amount of space available to us for information storage (or other uses) is enormous. As first described in a lecture titled, 'There's Plenty of Room at the Bottom' in 1959 by Richard P. Feynman, there is nothing besides our clumsy size that keeps us from using this space. In his time, it was not possible for us to manipulate single atoms or molecules because they were far too

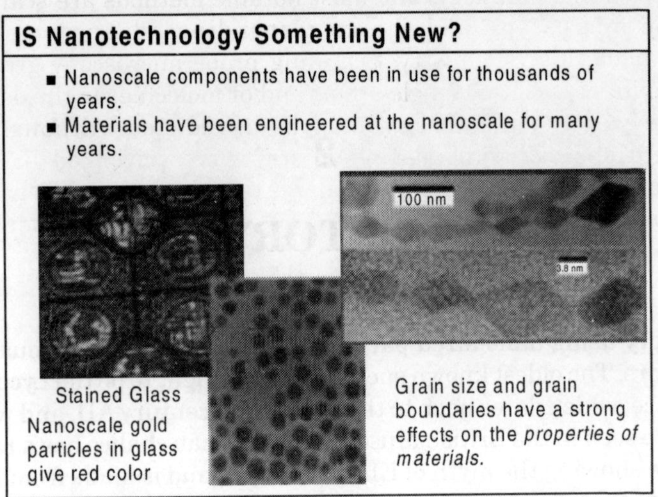

small for our tools. Thus, his speech was completely theoretical and seemingly fantastic. He described how the laws of physics do not limit our ability to manipulate single atoms and molecules. Instead, it was our lack of the appropriate methods for doing so. However, he correctly predicted that the time would come in which atomically precise manipulation of matter would inevitably arrive.

Prof. Feynman described such atomic scale fabrication as a bottom-up approach, as opposed to the top-down approach (we will learn the in core concepts chapter) that we are accustomed to. The current top-down method for manufacturing involves the construction of parts through methods such as cutting, carving and molding. Using these methods, we have been able to fabricate a remarkable variety of machinery and electronics devices. However, the sizes at which we can make these devices is severely limited by our ability to cut, carve and mold.

Bottom-up manufacturing, on the other hand, would provide components made of single molecules, which are held together by covalent forces that are far stronger than the forces that hold together macro-scale components. Furthermore, the amount of information that could be stored in devices build from the bottom-up would be enormous.

Since that initial preview of nanotechnology, we have developed several methods which prove that Prof. Feynman was correct in his prophesy. The most notable methods are scanning probe microscopy and the corresponding advancements in supramolecular chemistry. Scanning probe microscopy gives us the ability to position single atoms and/or molecules in the desired place exactly as Prof. Feynman had predicted. Although the limitations of traditional chemistry were criticized in Prof. Feynman's lecture due to its seemingly tedious and random nature, recent advancements have improved its potential uses for nanotechnology.

Now we will learn how nanotechnology first impacted human lives and about the advances since then that have enabled our understanding and opened the doors to potential new discoveries.

PRE-18th CENTURY

600 B.C. Kanaad

Indian philosopher and saint conceived of matter as being composed of smallest individual discrete particles. He called them "paramannu". He didn't tell anyone about 'Paramaanu' due to fear of misuse.

Roman Period (30 BC-640 AD) (Lycurgus Cup)

Archeological remains give clues to the use of nanotechnology materials in ancient times. A famous artifact from this period called the Lycurgus cup resides in the British Museum in London. The base of the Lycurgus Cup is made from glass and dates from the fourth century AD (the gilded bronze base and rim were added later). What makes this cup unique is that its color changes from green (when illuminated from the outside) to red (when illuminated from within). What causes the color change? Transmission electron microscopy reveals that the glass contains nanoparticles of gold and silver. At the nanoscale, materials exhibit properties that are differnt from their macroscale counterparts. Most likely the unique properties of this ancient Roman piece were created by accident as there are surviving pieces from this era that appear to be failed attempts to recreate this effect.

Medieval Period (500-1450 AD)

Although unaware of the reasons, stained glass artists were early users of nanotechnology.

The ruby red color of some stained glass is due to gold nanoparticles trapped in the glass matrix, while the deep yellow color is due to silver nanoparticles. The size of the metal nanoparticles produced these color variations. This example of the dramatic change in material properties (in this case, color) at the nanoscale is a key component of nanotechnology.

Renaissance Period (1450-1600 AD)

Artisans coloring pottery in fifteenth-and sixteenth century Deruta Umbria were practicing an early from of nanotechnology. Deruta ceramicists produced dramatic indescent or metallic glazes, which during the fifteenth and sixteenth centuries were in demand throughout Europe. To achieve the red and gold luster effects, particles of copper and silver metal between 5 and 100 biliionths of a meter were used. Instead of scattering light, the particles cause light to bounce off their surface at different wavelengths, giving metallic or iridescent effects.

During the same period in India many artisans had created wonderful objects based on nanotech that are placed now in Jodhpur, Udaipur, Bundi museum.

19th CENTURY

1803 (John Dalton)

- Chemical elements are made of atoms.
- The atoms of an element are identical in their masses.
- Atoms of different elements have different masses.
- Atoms only combine in small, whole number ratios such as 1:1, 1:2, 2:3 and so on.
- Atoms can be neither created nor destroyed.

1827 (Photography)

Photography is an early example of nanotechnology, which depends on the production of silver nanoparticles sensitive to light. Photographic film is a thin layer of gelatin containing silver halides and a base of transparent cellulose acetate. The light decomposes the silver halides, producing nanoparticles of silver, which are the pixels of the photographic image. In the late eighteenth century, the British scientists Thomas Wedgewood and Sir Humphry Davy were able to produce images using silver nitrate and chloride, but their images were not permanent. The first successful photograph was produced in 1827, by Joseph Niepce using material that hardened on exposure to light. This picture required an exposure of eight hours. Niepce went into partnership with Louis Daguerre. Although Niepce died from a stroke only four years later, Daguerr continued to experiment and in 1839 discovered a way of developing photographic plates, a process which greatly reduced the exposure time from eight hours to half an hour. He also discovered that an image could be made permanent by immersing it in salt.

1857 (Discovery of Gold Colloids)

Althoug the term "nano" was not in use at the time, early researcher Michael Faraday discovered and prepared the first metallic colloids in 1856. Colloids are fine particles that suspend in a solution (in between particles that dissolve in solution and those that settle). Faraday's gold colloids had special electronic and optical properties, and are now known as one of the many interesting metallic nanoparticles.

Considered by many to be one of the greatest experimentalists who ever lived, the English chemist and physicist Faraday received little more than a primary education, and at the age of 14 was apprenticed to a bookbinder. There he became interested in the physical and chemical works of the time. After hearing a lecture by the famous chemist Humphrey Davy, he sent Davy the notes he had made of his lectures. As a result Faraday was appointed assistant to Davy in the laboratory of the Royal Institution in London at the age of 21. The entire set of specimens (over 600) used by Faraday in his 1856 research are still housed at the Roay Institution of Great Britain.

20th CENTURY

1908 (MIE Theory)

German physicist, Gustav Mie played a hand in nanotechnology with his theory of light scattering by particles. His theory shows that light scatters from particles more efficiently at short wavelengths than at long wavelengths. For example, we see the sky as being blue because the molecules in air (which are extremely small particles) scatter light from the sun more efficiently for blue light than for yellow or red, as blue light has the shorter wavelength.

When the sun sets, the sunlight travels through the atmosphere over a longer distance than when it is overhead. The most important scattering in this case arises from dust particles. These particles will scatter light more effectively for blue colors, so the light that is not scattered is a mixture of red and yellow. This produces the characteristic red color of the setting sun.

Mie theory helped scientists to realize that the size of particles determines the colors that we see. Mie went on to develop a way to calculate the size of particles by determining the light they scatter. For nanoparticles and larger particles, this theory requires a huge number of calculations. so it was rarely used until about 20 years ago when supercomputers became available. Now, Mie theory (as well as other developed more recently) helps researchers predict and determine the size of nanoparticles.

1931 (Electron Microscope)

Although light microscopes had been around since the Renaissance, they were unable to recognize objects that were smaller than the wavelength of visible light (0.4-0.7 micrometers). To see particles smaller than this scientists had to bypass light altogether and use a different sort of "illumination", one with a shorter wavelength.

In 1931, German scientists Max Knott and Ernst Ruska developed an entirely new type of microscope that would eventually open up a new "small" world. With the electron microscope, electrons are accelerated in a vacuum until their wavelength is extremely short, only one hundred-thousandth that of white light. Beams of these fast-moving electrons are focused on a sample. Parts of the sample soak up the electrons, other parts scatter them. An electron-sensitive photographic plate "records" this action and creates an image. In 1933 the electron microscope was able to exceed the detail and clarity of the traditional light microscope.This was an important first step in the development of techniques and instrumentation that would eventually enable research at the nanoscale.

1947 (The Transistor)

Until the mid-1940's vacuum tubes were the state of the art in electronics. Capable of converting alternating current to direct current (AC to DC) andamplifying an electronic signal, vacuum tubes were used in everything from switching telephone calls to building the first high-speed computer, ENIAC. But the limitations were clear. The tubes were bulky and to make more powerful computers more tubes were needed (17,000 tubes were used in ENIAC). The tubes were also fragile and overheated easily. In 1945, Bell Labs established a research group to look into finding a solution. The group was led by Williams Shockley and included Walter Brattain and John Bardeen.

After two years, Bardeen and Brattain created an amplifying circuit that seemed to work, using the element germanium. They called it the point-contact transistor. The discovery did not gain attention until 1951, when Shockley improved upon the original idea with a junction transistor. The transistor was a solid (giving rise to the term "solid-state technology"), but had the electrical properties of a vacuum tube. Furthermore, it was inexpensive, slurdy, used little power, worked instantly, and best of all, was tiny. The three men shared the 1956 Nobel Prize in physics "for their researches on semiconductors and their discovery of the transistor effect." The invention of the transistor and the intergrated circuit marked the beginning of microelectronics, a field that relies on tools for miniaturization. The semiconductor industry is one of the largest technology drivers in the field of nanotechnology. Researches today are looking to enable the creation of chips holding billions or even trillions of nanoscale transistors.

1951 (Field-ion electron Microscope)

The development of nanotechnology was, and is, dependent upon advances in scientists instrumentation. Erwin Mueller, a professor at Penn State University's, made an important contribution when he invented the field-ion electron microscope in 1951. For the first time in history, individual atoms and their arrangement on a surface could be seen. For this accomplishment. Professor Mueller is known as the first person to "see" atoms. The device was a landmark advance in scientific instrumentation that allowed a magnification of more than 2 million times.

1953 (Discover of DNA)

One of the landmark achievements in the 20th century was the discovery of DNA. By the 1950s, scientists already knew that DNA-deoxyribonucleic acid-carried genetic information, but they didn't know what it looked like or how it worked.

In 1953, Dr. James Watson and Professor Francis Crick published an article in Nature describing the double helix structure of DNA. The showed that when cells divide, the two strands that make up the DNA helix separate and a new "other half" is built on each strand, a copy of the one before. This means that DNA can reproduce itself without changing its structure, except for occasional errors or "mutations." James Watson. Francis Crick.

Decades later, DNA's ability to self-assemble into tiny structures would inspire researches to use the same principles to develop nanoscale structures with specific dimensions and chemical properties.

1958 (Tunneling Phenomenon)

In 1958, Leo Esaki, a Japanese physicist working at Sony Corporation, discovered that electrons could sometimes "tunnel" through a potential barrier formed at the junctions of certain semiconductors even though classical theory predicted that this was not possible. What Dr. Esaki observed was an example of how materials at the nanoscale are controlled by different laws of behavior, i.e. they are controlled by quantum mechanics as opposed to classical physics. The discovery lead to the creation of the tunneling diode (sometimes called the Esaki diode), an improtant component of solidstate physics and the first time that tunneling (an important nano-electronics phenomenon) was used in a real device.

1959 (Richard Feynman)

Richard P. Feynman is often credited with predicting the possibilities and potentials of nano-sized materials. Feynman, who would go on to win the Nobel Prize in physics, gave a talk on December 29, 1959 entitled. "There's Plenty of Room at the Bottom." In his speed he stated.

"What I want to talk about is the problem of manipulating and controlling things on a small scale.

As soon as I mention this, peopole tell about miniaturization, and how far it has progressed today.

They tell me about electric motors that are the size of the nail on your small finger. And there is a device on the market, they tell me, by which you can write the Lord's Prayer on the head of a pin. But that's nothing; that's the most primitive, halting step in the direction I intend to discuss. It is a staggeringly small world that is below. In the year 2000, when they look back at this age, they will wonder why it was not until the year 1960 that anybody began seriously to move in this direction."

1960 (Ferrofluids)

In the 1960s, NASA researchers were trying to find ways to control liquids in space. They discovered that nano-sized magnetic particles of iron that were given a chemical coating or surfactant, that prevented them from clumping together could be dispersed in oil or water. They could then control the location of the fluid (called a ferrofluid) with a magnet.

On Earth, ferrofluids are used inside loudspeakers, where they help keep the inner parts cool. They are also used on computer hard drives and in semiconductor manufacturing, as seals to keep out dust and other contaminants.

Nanotechnologies want to harness ferrofluids for other important uses such as developing tiny sensors, or inside the body as biomedical devices, to deliver drugs, absorb toxins, or hypothermia. It is even possible that ferrofluids could be used to help clean up hazardous waste spills.

1960 (Zeolite Catalysts)

A zerolite is an inorganic porous material which works as a kind of molecular sieve-allowing some molecules to pass through while excluding or breaking down others. Zeolites can be natural or synthetic, and new zeolites are still being discovered and invented.

In 1960, Charles Plank and Edward Rosinski developed a process to use zeolites to speed up chemical reactions. Plank and Rosinski's process focused on using zeolites to break down petroleum into gasoline more quickly and efficiently.

Researchers are currently working to design zeolite catalyst at the nanoscale. By adjusting the size of the zeolite pores on the nanoscale, they can control the size and shape of molecules that can enter. In the case of gasoline production, this technique could mean that we would get more and cleaner gasoline from every barrel of oil. It is hoped that nanotechnology will help reduce the cost and pollution associated with producing gasoline and other petroleum products.

1965 (Moore's law)

Writing for Electronics magazine in 1965, Gordon E. Moore, the founder of Intel Corporation, noted that the number of transistors per integrated circuit has doubled every two years. He predicted that this trend would continue for another 10 years and his prediction was quickly dubbed "Moore's law" by the press. Moore's prediction turned out to be prophetic. In fact, the complexity of a chip continued to double yearly, long after 1975. The rate of doubling has only recently slowed to about every 18 months. Many researchers believe that devices which use electronic nanotechnology and molecular electronics will keep Moore's law accurate into the future.

1970 (Sir John Pople's Software)

Developing new mathematical formulas is an essential part of science, and as science becomes more complicated, new, faster formulas are needed. Although computers had become dramatically more powerful by this time, new software must be developed in order for that power to be useful. Up until this time, it was very difficult to solve the complex mathematical equations needed to determine the properties of molecules. In 1970, John Pople and his research group developed Gaussian, a software program that would preform these calculations. This pioneer in the use of computers to predict the behavior of atoms and molecules also developed many of the algorithms that made computer-based modeling at the nanoscale possible.

1974 (Term "Nanotechnology" First Used)

Ther term "Nanotechnology" was first used by Norio Taniguchi of the Tokyo Science University. He used the word to refer to "production technology to get the extra high accuracy and ultra fine dimensions, i.e. the preciseness and fineness on the order of 1 nanometer."

1974 (Molecular Electronics)

In 1974, Northwestern University Charles and Emma Morrison Professor of Chemistry, Mark A. Ratner and A. Aviram of IBM, proposed that individual molecules might exhibit the behavior of basic electronic devices, thus allowing computers to be built from the bottom up by turning individual molecules into circuit components. This hypothetical application of nanotechnology, formulated long before the means existed to test it, was so radical that it wasn't pursued or even widely understood for another 15 years. For this groundbreaking work, Ratner is widely credited as the "father of molecular-scale electronic" and his contributions were recognized in 2001 with the Feynman Prize in Nanotechnology.

1977 [Surface Enhanced Raman Spectroscopy (SER)]

Some of the tools needed in the fied of nanotechnology were many years in the making. For example, spectroscopy is a set of techniques that use the interaction of light with matter to otbain information about its density and structure of molecules. Sir Chandrasejgara Venkata Raman (Calcutta University) won the 1930 Nobel Prize in physics for his 1928 discovery that the scattering of light by molecules could be used to provide information about a sample's chemical composition and molecular structure. Although a ground-breaking technique. Raman Spectroscopy was not capable of detecting at the nanoscale. The technique was greatly improved in the 1960s by the invention of the laser, but it wasn't until 1977, when Richard P. Van Duyne discovered Surface Enhanced Raman spectroscopy (SERS), that nanoscale studies became possible. Van Duyne deduced that when molecules were attached to a surface which had hills and valleys that were approximately 50-100 nanometers in size, the Raman intensity was amplified 1,000,000 times. The discovery of SERS completely transformed Raman Spectroscopy from one of the least sensitive to one of the most sensitive techniques in all of molecular spectroscopy. Today, SERS is used to study the chemical reactions of molecules in electrochemistry, catalysis, materials synthesis, and biochemistry. The sensitivity of SERS is now so high that even single molecules can be studied.

1980 [Self-Assembled Monolayers (SAMs)]

In 1980, Jacob Sagiv at the Weitzman Institute in Israel discovered that molecules containing a chemcial called octadecyltrichlorosilane or OTS would spontaneously react with a glass surface to assemble by themselves into individual layers. In 1983, a Bell Labs research team lead by David Allara discovered that molecules with thiol groups (groups containing sulfur) on a gold surface would also self-assemble into individual or nano-layers. These self-assembled monolayers are typically a few nanometers thick (determined by the choice of molecule) and allow researchers to tailor the properties of a surface. For the first time, scientists could envision building three dimensional nanoscale structures layer-by-layer. Like laying rows of bricks to build a wall. These structures are being used to build molecule-based electronic devices, biosensors, and new types of optical materials.

1981 [Scanning tunneling microscope (STM)]

In 1981, the scanning tunneling microscope (STM) was invented by Gerd Bining and Heinrich Rohrer at IBM's Research Laboratory in Zurich, Switzerland. This invention allowed scientists not only to observe nanoscale particles, atoms, and small molecules, but to control them. The STM scans the tip of a needs, or "probe", just a few atoms above the surface of the sample. A voltage is applied between the tip of the probe and the surface. As the electricity begins to flow, the STM can determine minute variations in the distance electrons travel. In this way, the STM maps the surface the sample. The information is saved in a data file and a "picture" of the surface is created by computer. In this way, the STM can "see" atomic-scale objects. The STM helps researchers determine the size and form of molecules, observe defects and abnormalities, and discovere how chemicals interact with the sample. The STM quickly became standard equipment in laboratories thorugh the world.

1985 (The "Buckyball")

Another nanotechnology breakthrough occured in 1985, when Richard Smalley, Robert Curl and graduate student James Heath at Rice University, and Sir Harry Kroto at the University of Sussex discovered C60, a carbon nanoparticle shaped like a soccor ball. The unique molecule, was named Buckministerfullerene after the visionary America architect and enginner Buckminster Fuller who designed the geodesic dome. More commonly called a "Buckyball," the molecule is extremely rugged capable of surviving collisions with metals and other materials at speeds higher than 20,000 miles per hour. Because of this ruggedness, buckyballs show promise in the development of fuel cells that might power the automobiles of the future. Researchers are also investigating the possibility of using buckyballs as tiny drug delivery systems.

1986 (Atomic Force Microscope)

Invented by Gerg Binnig and his colleague Christoph Gerber at IBM, San Jose, and Calvin Quate at Stanford University, the atomic force microscope or AFM uses a cantilever to "read" a surface directly, the way a record player's needle reads a record. Atomic force microscopy works by passing the cantilever—so sharp that its tip is composed of a single atom—within a few nanometers of a surface. The atomic forces exerting a pull on the cantilever are measured to create an atom-by-atom topographical map. The AFM makes 3-D images of an object's surface topography with extremely high magnifications (up to 1,000,000 times).

1987 (Single-Electron Tunneling Transistor)

In 1985 Dmitri Averin and Konstantin Likharev, then at the University of Moscow, proposed the idea of a new device called a single-electron tunneling (SET) transistor. Two years later Theodore Fulton and Gerald Dolan at Bell Labs in the US built such a device. In this structure the controlled movement of individual electrons through a nanoscale device was first achieved.

Single-electron devices are based on what is called the tunnel effect. When two metallic electrodes are separated by an insulating barrier about 1 nanometer thick (approximately 3 atoms in a row), the electrons are able to "tunnel" through the insulator, even though classical theory suggests that this is impossible.

Researchers have long considered whether SET transistors could be used for digital electronics, but the random variations in voltage from device to device caused serious problems. Working at the nanoscale, today's researchers are considering how to overcome this problem by combining all of the components of the SET transisitor into a single molecule. It is possible that conventional circuits will one day be replaced by electronics based upon individual molecules and form the basis for a new class of nanoelectronics.

1988 (Discovery of Quantum Dots)

In early 1980s, Dr. Louis Brus and his team of researchers at Bell Laboraotories made a singnificant contribution to the field of nanotechnology when they discovered that nano-sized crystal semiconductor materials made from the same substance exhibited strikingly different colors. These nanocrystal semiconductors were called quantum dots and this work eventually contributed to the understanding of the Quantum Confinement Effect, which explains the relationship between size and color for these nanocrystals.

Due to their extraordinary small size, the electrons inside the quantum dots exhibit unique behavior. Specificially, the electrons are confined to far fewer energy levels than allowed in bulk semiconductor materials. This results in the quantum dots emitting very intense light of a specific color when the electrons make transitions between these discrete energy levels. Small differences in the size of the quantum dot change the allowed electron energies and therefore alter the color light which they emit. Scientists have learned how to control the size of quantum dots making it possible to obtain a broad range of colors.

Quantum dots have the potential to revolutionize the way solar energy is collected, improve medical diagnostics by providing efficient biological markers, and advance the development of optical devices such as light emitting diodes (LEDs).

1990 (Manipulation of Atoms)

Using a scanning tunneling microscope (STM), IBM researchers Donald Eigler and Erhard Schweizer were able to arrange individual Xe atoms on a surface. Although the process was painstakingly slow, the remarkable image allowed researchers to place individual xenon atoms with nanoscale precision and to visualize the results. This now famous image of the atomic world "hangs" in IBM's STM Image Gallery, and demonstrates early attempts to create structures one atom at a time.

1991 (Carbon Nanotubes)

In 1991, Sumi Lijima at NEC in Japan discovered a new form of carbon called nanotubes, which consisted of several tubes nested inside each other. Two years later lijima, Donald Bethune at IBM in the US and others observed single-walled nanotubes just 1-2 nanometers in diameter.

Nanotubes behave like metals or semiconductors, but can conduct electricity better than copper, can transmit heat better than diamond, and are among the strongest materials known. Nanotubes could play a pivotal role in the practical applications of nanotechnology if their remarkable electrical and mechanical properties can be exploited.

1996 (Using DNA and Gold Colloids to Assemble Inorganic Materials)

Since optical properties of gold colloids in 1857, researchers have sought to harness these capabilities. In 1996, Northwestern University researchers, Chad Mirkin and Robert Letsinger, discovered a way to do this. They attached strands of synthetic DNA onto gold nanoparticles. Since complementary strands of DNA can recognize and bind to each other, the DNA served as a blueprint, a construction worker, and a sorter to create new inorganic materials. By manipulating the DNA, they were able to make materials with the same unusual properties as the nanoscale building blocks that they are made from. This advance generated an explosion of interest in making designer bio-inorganic architectures at the nanoscale.

1999 (Development of Dip-Pen Nanotechnology)

A pivotal development in the constellation of nanotechnology tools was Dip-Pen Nanlithography or DPN. Invented in 1999 by Chad A. Mirkin. The concept is based upon a classic quill pen—a 4,000 year old technology. Using an atomic microscope tip. DPN allows researchers to precisely lay down or "write" chemicals, metals, biological macromolecules, and other molecular "inks" with nanometer dimensions and precision on a surface. DPN has progressed to include 1,000,000 tip serial and parallel processing - opening the door to credible nano-manufacturing techniques for smaller, lighter weight, faster, and more reliably produced electronic circuits and devices, high-density storage materials, and biological and chemical sensors.

21st CENTURY

2000 [Feedback-controlled Lithography (FCL)]

Feedback-controlled Lithography (FCL) is a technique that allows researchers to use the Scanning Tunneling Microscoope (STM) to precisely and selectively build structures at the nanoscale.

Developed by Mark Hersam and Joseph Lyding, FCL is conducted by first coating a silicon sample with hydrogen (otherwise known as hydrogen passivated silicon). Using the STM, researchers can then image (or see) the surface of the silicon sample. When electric voltage is applied to the STM tip from an outside source, the silicon-hydrogen bounds are broken. By controlling the position of the STM tip, hydrogen atoms are thus removed with atomic precision.

This technique allows fundamental studies of chemistry at the single molecule level and has opened the door for building prototype devices and other structures at the nanoscale.

2002

1. *Chipmaking Made Easier: Laser Assisted Direct Imprinting (LADI)*

LADI could allow electronics manufacturers to increase the density of transistors on silicon chips 100-fold.What's more, traditional photolithographic etching takes 10 or 20 minutes to make a single chip. Chou's imprint method accomplishes that in a quarter of a millionth of a second.

2. *Better than Silicon*

Single-Wall Carbon Nanotube Field Effect Transistors team at IBM prove that carbon nanotubes outperform silicon transistor prototypes available today, but he also developed a process to make the goal of a commercially available product more attainable. Metals have traditionally been used to help the nanotubes form, but end up contaminating the final product because they create soot which then must be filtered out. But Avouris developed a way to get around this and to control the alignment of nanotubes so that transistor arrays can be constructed. The nanotube transistor produced offered more than twice the current carrying capability per unit width of top-performing silicon transistor prototypes.

3. *Nanoscale Soldering*

Nanowire Two different materials can be placed next to one another on the same-nanometer sized crystal in a manner that creates a nanowire with a defect-free electrical junction. This capability could allow Lieber's nanowires to be the building blocks for almost any kind of applications in electronics, photonics or bio-chemical detection.

4. *Moving Toward a Molecular Future*

HP and IBM's Advances in Molecular ElectronicsBy manipulating individual atoms and molecules to form logic gates and memory circuits, researchers at Hewlett-Packard [HPQ] and IBM amazed even their peers. Logic circuits created out of molecules are 260,000 times smaller than today's silicon-based chips. HP unveiled a simple, inexpensive and scalable chemical process that could be used in creating a variety of molecular-scale electronic devices including logic, memory, communications and signal routing devices IBM created the world's smallest logic circuit using a technique called "molecular cascading."

5. Super-Dense Hard Drives

Ballistic Magnetoresistance Working with a sensor 40nm wide, SUNYBuffalo's Harsh Chopra made a major discovery that could have profound effects on the magnetic storage industry. As electrons flow through materials like metal, they ordinarily scatter, bouncing in different directions. An electrical current arises when the movement of electrons is a flow in one direction. In wires only a few nanometers wide, Chopra saw that the scattering stops and electrons flow straight through, or ballistically. Because of the nanoscale size of the sensor, it would be possible to produce a CD-sized magnetic disk that could hold 1,800 gigabytes—about 450 DVDs worth of information.

2003

IBM demonstrated new nanotechnology method to build chip components IBM announced it was the first to successfully apply a novel approach in nanotechnology to aid conventional semiconductor processing, potentially enabling continued device miniaturization and chip performance improvements. IBM used a "molecular self-assembly" technique that is compatible with existing chip-making tools, making it attractive for applications in future microelectronics technologies because it avoids the high cost of tooling changes and the risks associated with major process changes.

IBM's self-assembly technique relies on the tendency of certain types of polymer molecules to organize themselves. The polymer molecules pattern critical device features that are smaller, denser, more precise and more uniform than can be achieved using conventional methods like lithography. The use of techniques such as self assembly could ultimately lead to more powerful electronic devices such as microprocessors used in the growing array of computer systems, communications devices and consumer electronics. IBM expects self-assembly techniques could be used in pilot phases in three to five years.

IBM and Infineon presented their high-speed 128Kbit MRAM core at the VLSI Symposia in Kyoto, Japan. It is fabricated with a 0.18 micron logic-based process technology, the smallest size reported to date for MRAM technology. This small base enabled the two companies to incorporate the smallest MRAM memory-cell size of 1.4 square microns, which is about 20 million times smaller than the average pencil eraser top. By accurately patterning the magnetic structures within this small cell, researchers were able to control the memory reading and writing operations.

A memory technology that uses magnetic, rather than electronic, charges to store bits of data, MRAM could significantly improve portable computing products by storing more information, accessing it fater and using less battery power than the electronic memory used today. MRAM combines the best features of today's common memory technologies: the storage capacity and low cost of Dynamic RAM (DRAM), the high speed of Static RAM (SRAM) and the non-volatility of flash memory. Since MRAM retains information when power is turned off, products (personal computers, for instance) using it could start up instantly, without waiting for software to boot up.

2004/2005

Nanotechnology Toolbox

A new way to make structures from folded RNA has been discovered. Nucleic acids previously have been used to make two dimensional arrays and three dimensional polygons. Now it appears they can make three dimensional arrays as well.

This is another tool in the toolbox of ways to make large structures out of molecules. It's worth noting that the researchers talk about "incorporating these nanomachines into nanodevices."

Toshiba Develops MEMS Based Manipulation Technology for Injecting Nanoparticles in Cells.

Compared with conventional techniques using laser beam to affect cells physically, the newly developed technology has advantages, such as simultaneous manipulation of numerous cells. Expected applications in the field of biotechnology include a medical analytical tool for investigating the reaction of cells to physical effects and clarifying their detailed properties, and, looking further ahead, a technique for affecting specific cells.

Nano-based Antiradiation Drug

The researchers found that buckyballs given before or immediately after exposure to X-rays reduced organ damage by one-half to two-thirds, which is as good as the level of protection amifostine provides.

Dicker noted C Sixty can coat the buckyballs with molecules that could allow the nanoparticles to target and protect specific organs and tissues from radiation

Engineers make Standardized Bulk Synthesis of Nanowires possible graduate student Eric Stern in the department of biomedical engineering along with his colleague Guosheng Cheng, associate research scientist in electrical engineering systematically varied and tested parameters for producing GaN NWs using an optical lithographic method as a template for testing characteristics of the NWs

Development of reliable NW fabrication will allow the exploration of the next steps in semiconductor miniaturization. This reported technology produces ten-times the number of NWs as previous technology and sets parameters for standardization of NWs.

SO, WHAT IS NEW?

- We now have the ability to construct by synthesizing from atoms up (bottom-up) and pattern down from larger assemblies (top-down). We have sophisticated tools to characterize the nanoscale.
- We can explore, characterize, and utilize across a breadth of the disciplines.
- Nanoscale has brought together approaches of different disciplines-physical and life sciences, and enginnering -

at the length scale of the building blocks of the physical and living world.

- The opportunities of understanding and in turn benefiting humanity by harnessing these capabilities is enormous
  - ➤ The scientific and engineering potential is vast.
  - ➤ The interdisicplinary research and applications are exciting.

3

CORE CONCEPT OF NANOTECHNOLOGY

1. NANOTECHNOLOGY

(i) A technology that operates at the atomic, molecular or macromolecular levels, in a length scale of 1-100 nanometers.

(ii) A technology that creates or uses structures, devices and systems that have novel properties and functions because they harness forces that are only evident at level of atoms and molecules (the nanoscale)—in other words, technologies that are useful because of properties or capabilities that only a nanoscale device could offer.

(iii) Technologies that control or manipulate matter at the atomic scale—technologies that are capable of manipulating atoms and molecules by design.

As things approach the nanoscale, new properties emerge due to size confinement, quantum phenomena, and coulomb blockage. These new properties can be controlled to give us materials with new applications. Specifically, nanotechnology will permit control of the following:

- Structural properties (e.g. strength and ductility)
- Electrical properties
- Magnetic properties
- Catalytic properties
- Thermal properties
- Optical properties
- Biocompatibility

These properties can then be brought to the macro world through one of several fabrication processes:

- Lithography (e-beam, x-ray, ion beam, soft/nanoimprint, dip pen)
- Sol-gel
- Chemical Vapor Deposition
- Hierarchical templated assembly
- Self-assembly/Biologically assisted assembly (bacterial, viral, etc.)

Nanotechnology could be...

- Nano (metrology)
- Nanoscale science (effects)
- Nanoscale technology (fabrication)
- Molecular nanotechnology (chemistry)

Nanotech Generation

First Generation

Passive nanostructures in coatings, nanoparticles, bulk materials (nanostructured metals, polymers, ceramics): ~2001

Second Generation

Active nanostructures such as transistors, amplifiers, actuators, adaptive structures: ~2005

Third Generation

3D nanosystems with heterogeneous nanocomponents and various assembling techniques: ~2010

Fourth Generation

Molecular nanosystems with heterogeneous molecules, based on biomimetics and new design ~2020 (?)

2. NANOSCIENCE

It was defined as 'the study of phenomena and manipulation of materials' at nanoscale-research which will fuel technology advances just around the corner.

We should distinguish between nanoscience, which is here now and flourishing, and nanotechnology, which is still in its infancy. Nanoscience is a convergence of physics, chemistry, materials science and biology, which deals with the manipulation and characterisation of matter on length scales between the molecular and the micron size. Nanotechnology is an emerging engineering discipline that applies methods from nanoscience to create products.

What is Special About Nanoscience?

The laws of physics operate in unfamiliar ways on these length scales, and this is important to appreciate for two reasons.The peculiarities in behaviour imposed by the nanoscale impose strong constraints on what is possible to design and make on this scale. But the very different behaviour of matter on the nanoscale also offers opportunities for structures and devices that operate on radically different principles from those that underlie the operation of familiar macroscopic objects and devices. For example, the importance of quantum effects could lead to highly novel computer architectures—quantum computing—while the importance of Brownian motion and surface forces leads to an entirely different principle for constructing structures and devices—self-assembly. Key differences in the way physics operates at the nanoscale include:

Quantum Physics

On small length scales matter behaves in a way that respects the laws of quantum mechanics, rather than the familiar Newtonian mechanics that operates in the macroscopic world.These effects are particularly important for electrons. One example arises from Heisenberg's uncertainty principle, which states that we cannot know accurately and simultaneously the position and momentum of a particle.

If we confine an electron by reducing the dimensions of a metal or semiconductor particle, then its energy has to increase, in effect to compensate for its spatial localisation. This means that confinement can be used to modify the energy levels of electrons in semiconductors, to create novel materials whose optoelectronic properties can be designed to order.

Brownian Motion

Submicron particles and structures immersed in water are subject to continuous bombardment from the molecules around them, causing them to move about and internally flex in a random and uncontrollable way. If we expect nanomachines to work according to the principles of macroscopic engineering, Brownian motion imposes strong constraints on the stiffness of the component materials and the operating temperatures of the device. In the view of many scientists this renders impractical some radical proposals for nanodevices which consist of assemblies of molecular-scale cogs and gears. On the other hand, some biological nanodevices, like molecular motors, are clearly not subject to these constraints, because their mode of operation actually depends in a deep way on Brownian motion.

Surface Forces

Surfaces and interfaces play an increasingly important role for particles or structures as they are made smaller. A variety of physical mechanisms underlie the forces that act at surfaces (at a macroscopic scale, the surface tension that allows a water beetle to walk on water is an example of one of these), but the overall effect is simple; small objects have a very strong tendency to stick together.This stickiness at the nanoscale, and the accompanying strong friction that occurs when parts are made to move against each other, are an important factor limiting the degree to which microelectronic mechanical systems (MEMS) technologies can be scaled down to the nanoscale.These phenomena also underlie the almost universal tendency of protein molecules to stick to any surface immersed within the body, with important consequences for the design of biomedical nanodevices.

Although the combination of Brownian motion and strong surface forces is sometimes thought of as a problem that nanotechnology must overcome, these features of the nanoworld in fact combine to offer a remarkable opportunity to exploit an approach to fabricating devices peculiar to the nanoscale. If molecules are synthesised with a certain pattern of sticky and non-sticky patches, the agitation provided by Brownian motion can lead to the molecules sticking together in well-defined ways to make rather complex nanoscale structures. The key to understanding this mode of assembly known as self-assembly is that all the information necessary to specify the structure is encoded in the structure of the molecules themselves. This is in contrast to the methods of directed assembly that we are familiar with at the macroscale, in which the object is built, whether by a tool-using human being or by a machine, according to some externally defined plan or blueprint. The attraction of self-assembly as a route to creating nanostructures is that it is parallel and scalable the number of structures created is limited only by how many molecules are put in. This is in contrast to the serial processes that are familiar at the macroscale, in which objects are created one at a time.

Self-assembly is an example of an approach to making nanostructures which is often referred to as 'bottom-up' nanotechnology. This term indicates approaches which start with small components—almost always individual molecules—which are assembled to make the desired structure. Bottom-up nanotechnology does not necessarily involve self-assembly. An alternative, but much less well developed, realisation of a bottom-up approach uses scanning probe microscopes to position reactive molecules at the desired position on surfaces.

In the opposite approach—'top-down' nanotechnology—one starts with a larger block of material and by physical methods carves out the desired nanostructure, as you would make a statue from a block of marble. Top-down nanotechnology is a natural extension of current methods of microelectronics, in which structures of very limited dimensions are created by laying down thin layers of material and etching away those parts of each layer that are unwanted.

The epitome of bottom-up processing technologies is provided by biology. Nanoscience is thought of as a physical science, but cell biology operates on exactly these length scales. The nanoscale devices that carry out the functions of living cells—the ribosomes that synthesise new proteins according to the blueprint provided by DNA, the chloroplasts that harvest the energy of light and convert it into chemical fuel, the molecular motors that move components around within cells and which in combination allow whole cells and indeed whole multicellular organisms to move around—are all precisely the kinds of machines imagined by nanotechnologists. Cell biology offers a proof that at least one kind of nanotechnology is possible.

Biology can provide lessons for nanotechnology. Longueur of evolution have allowed the perfection of devices optimised for working in the unfamiliar conditions that prevail at the nanoscale, and careful study of the mechanisms by which they work should suggest designs for synthetic analogues. This may lead to the design of synthetic molecular motors, selective valves and pores, and pumps that can move molecules around against concentration gradients.

3. MATERIALS SCIENCE

Materials science, the science of metals, ceramics, colloids and polymers, has always concerned itself with controlling the structure of materials on the nanoscale. Here nanoscale science and technology will largely facilitate incremental advances on existing materials and technologies. The improved control over nanoscale structure, and better understanding of relationships between structure and properties, will continue the long-run trend towards materials that are stronger and tougher for their weight.

Already, this is leading to reductions in the amount of material needed to make artefacts and thus, for example, to improved fuel efficiency in cars and aeroplanes. Control of structure on the nanoscale has been used for some time to improve the performance of magnetic materials, and this progress in turn will contribute to improvements in performance of electric motors and generators. Other types of functional materials—in particular those that are used in batteries and fuel cells—are also being improved in the same way, and the results, in terms of lighter and more efficient portable power sources, are being seen in devices such as mobile phones and laptop computers. In the control of surface properties in textiles and paints, improved materials are being developed with properties such as the breathability of waterproof fabrics and stain resistance in clothes and carpets.

4. NANOSCALE

Nanotechnology is the area of science and technology where *very very* small structures play a critical role, but how small is 'small'?

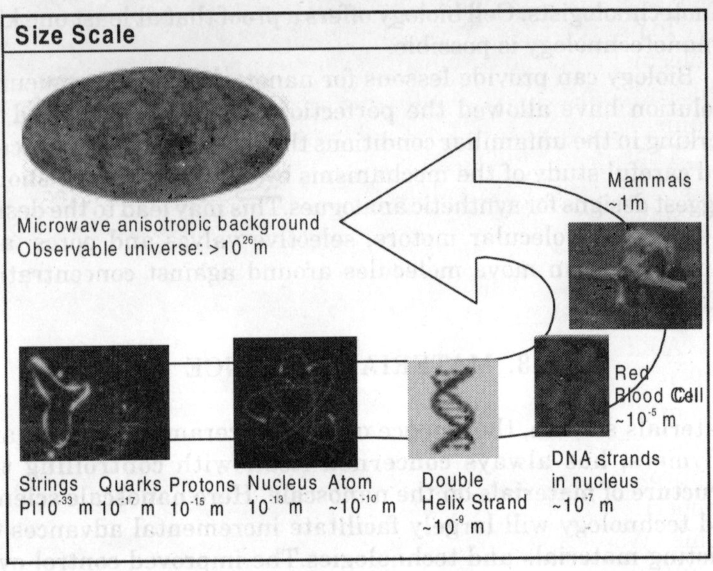

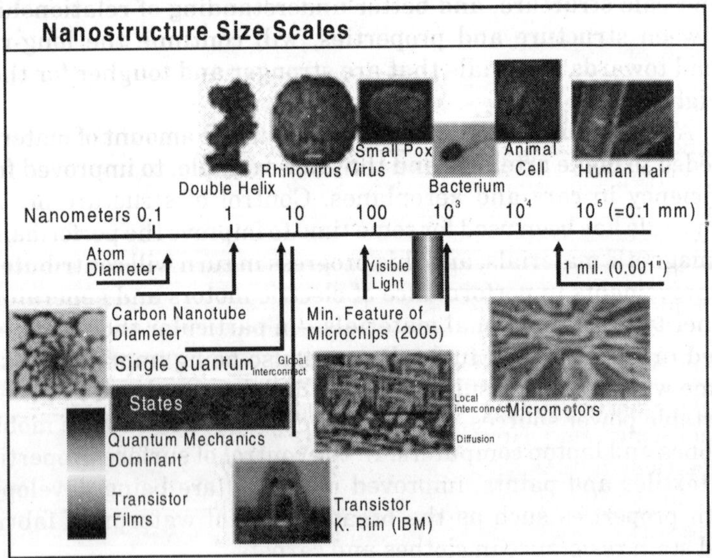

The scientific answer is that nanotechnology occurs at the order of nanometre (symbol: nm), that is from 0.1 nm to around 100 nm (which is 0.00000001 cm to 0.00001 cm).

A nanometer is defined as one billionth of a metre, which is a thousandth of a millionth of a metre, or 1×10^{-9} m.

To put it another way a nanometre is about as wide as a DNA molecule; how much your fingernails grow each second; 1/10 the thickness of the metal film on a packet of crisps; 1/80,000 the diameter of an average human hair.

That means that if a buckyball (60 carbon atoms arranged into a sphere with a diameter of approximately 1nm) were expanded to the size of a football, the football would corresponding be expanded so that it was much bigger than the size of Earth (becoming approximately the size of Neptune or Uranus about 50,000 km in diameter). So a nanometre is extremely small !

A single human hair is about 80,000 nm wide, a red blood cell is approximately 7,000 nm wide and a water molecule is almost 0.3nm across. People are interested in the nanoscale (which we define to be from 100nm down to the size of atoms (approximately 0.2nm)) because it is at this scale that the properties of materials can be very different from those at a larger scale. We define nanoscience as the study of phenomena and manipulation of materials at atomic, molecular and macromolecular scales, where properties differ significantly from those at a larger scale; and nanotechnologies as the design, characterisation, production and application of structures, devices and systems by controlling shape and size at the nanometre scale.

The properties of materials can be different at the nanoscale for two main reasons. First, nanomaterials have a relatively larger surface area when compared to the same mass of material produced in a larger form. This can make materials more chemically reactive (in some cases materials that are inert in their larger form are reactive when produced in their nanoscale form), and affect their strength or electrical properties. Second, quantum effects can begin to dominate the behaviour of matter at the nanoscale—particularly at the lower end—affecting the optical, electrical and magnetic behaviour of materials. Materials can be produced that are nanoscale in one dimension (for example, very thin surface coatings), in two dimensions (for example, nanowires and nanotubes) or in all three dimensions (for example, nanoparticles).

5. NEW FORMS OF CARBON

Carbon is one of the most familiar elements, well-known in its common form as graphite, in its rather rarer but much-prized form as diamond, and in various impure forms such as soot and charcoal.The discovery in 1985 of new forms of elemental carbon made a large impact and was rewarded with a Nobel Prize, and today forms the foundation for many hopes for nanotechnology.The reason is that these new forms of carbon are well-ordered structures that are intrinsically nanoscaled; Buckminster fullerene is a perfect sphere, made from exactly 60 carbon atoms. A good way of thinking of the fullerenes is as variants of graphite. Graphite consists of infinite flat sheets of carbon atoms in which each atom is linked to three other carbon atoms, so the pattern of bonds consists of tiled regular hexagons, exactly like a sheet of chicken wire. In fullerenes, some of the hexagons are replaced by pentagons, so that the resulting sheet is curved. In the case of C60 the pattern of hexagons and pentagons (which is that of a standard soccer ball) is such that the molecule is spherical.

Nanotubes are formed when the sheets of graphite are rolled up into tubes, the tubes being capped by fullerene hemispheres. Nanotubes can be as small as 2 nm in diameter, and can be perfectly regular in their arrangement of atoms.

A number of uses have been investigated or suggested for nanotubes.Very long nanotubes would be expected to be extremely strong and stiff, so if synthesis routes can be found to make them, they could potentially be used as ultra-strong, lightweight fibres. Even in shorter lengths, their mechanical properties should make them useful as reinforcing elements in composite materials. Nanotubes have useful electrical properties, being either electrically conducting or semiconducting, and have already been used to make nanoscale electronic devices. Nanotubes have already been used as tips for scanning force microscopes. Both nanotubes and fullerenes, such as C60, may find uses in new types of solar cells. A substantial effort has gone into devising routes to synthesise usefully large quantities of fullerenes and nanotubes, and we may expect the wider availability of the materials to lead to more intensive study and further applications.

6. NANOCOMPOSITES

The introduction of composite materials like glass- and carbon-reinforced plastics has led to new materials that have significantly higher performance for their weight than conventional ones. The benefits are now appearing in the aerospace sector. In these materials, a reinforcing material provides stiffness and strength while a much less stiff matrix material ensures toughness and reduces the weight. In current composites the reinforcing material is on a fairly large scale, and there are potential advantages if it could be made much smaller. The most popular realisation of this idea uses exfoliated clay platelets as the reinforcement, and applications of this technology in the automotive sector and the packaging industry are already with us. Many groups are working on using carbon nanotubes as a reinforcing material. To some extent this can be thought of as a development of carbon fibre reinforcement in which the fibres are particularly small and much more free from structural defects than the carbon fibres currently used. The resulting materials would be stronger, lighter, and stiffer than existing composite materials (such as carbon fibre reinforced plastics) and would find applications in the aerospace and automotive industries, if the price permitted.

Polymer Nanocomposites

Polyamides are widely used as engineering plastic materials for structural applications. In recent years several research groups have made attempts to improve the properties of polyamide 6 (PA6) by incorporation of unmodified or functionalized carbon nanotubes (CNT). An increase of 115 per cent in elastic modulus and of 124 per cent in tensile strength of PA6 have been reported by researchers in the melt mixed composites of PA6 and MWNT (purified and COOH and OH functionalized) at 1 wt% MWNT level. It is also reported that poly n-butyl acrylate (pBA) encapsulated MWNT when melt mixed with PA6 enhances interfacial adhesion between PA6 matrix and MWNT. This leads to the increased yield strength and tensile modulus of composites with 1 wt% modified MWNT as compared to pure PA6 and PA6/unmodified MWNT, whereas elongation at break ultimate tensile strength and impact strength decreases considerably. Indeed a vast amount of effort has been put up in the last few years in the field of CNT based polymer composites in order to obtain Polymer/CNT composites, which can be utilized as a multifunctional material.

7. NANOMATERIALS

Materials that consist of or contain nanoparticles, and can offer improved properties, such as lower weight, or higher strength.

We categorise nanomaterials as those which have structured components with at least one dimension less than 100nm. Materials that have one dimension in the nanoscale (and are extended in the other two dimensions) are layers, such as a thin films or surface coatings. Some of the features on computer chips come in this category. Materials that are nanoscale in two dimensions (and extended in one dimension) include nanowires and nanotubes. Materials that are nanoscale in three dimensions are particles, for example precipitates, colloids and quantum dots (tiny particles of semiconductor materials). Nanocrystalline materials, made up of nanometre-sized grains, also fall into this category. Some of these materials have been available for some time; others are genuinely new.

Two principal factors cause the properties of nanomaterials to differ significantly from other materials: increased relative surface area, and quantum effects. These factors can change or enhance properties such as reactivity, strength and electrical characteristics. As a particle decreases in size, a greater proportion of atoms are found at the surface compared to those inside. For example, a particle of size 30 nm has 5 per cent of its atoms on its surface, at 10 nm 20 per cent of its atoms, and at 3 nm 50 per cent of its atoms. Thus nanoparticles have a much greater surface area per unit mass compared with larger particles. As growth and catalytic chemical reactions occur at surfaces, this means that a given mass of material in nanoparticulate form will be much more reactive than the same mass of material made up of larger particles.

Properties of Nanomaterials

In tandem with surface-area effects, quantum effects can begin to dominate the properties of matter as size is reduced to the nanoscale. These can affect the optical, electrical and magnetic behaviour of materials, particularly as the structure or particle size approaches the smaller end of the nanoscale. Materials that exploit these effects include quantum dots, and quantum well lasers for optoelectronics.

For other materials such as crystalline solids, as the size of their structural components decreases, there is much greater interface area within the material; this can greatly affect both mechanical and electrical properties. For example, most metals are made up of small crystalline grains; the boundaries between the grain slow down or arrest the propagation of defects when the material is stressed, thus giving it strength. If these grains can be made very small, or even nanoscale in size, the interface area within the material greatly increases, which enhances its strength. For example, nanocrystalline nickel is as strong as hardened steel. Understanding surfaces and interfaces is a key challenge for those working on nanomaterials, and one where new imaging and analysis instruments are vital.

The Possible Toxicity of Nanoparticles

Given that many other properties of materials change when they are present in very finely divided form, it is reasonable to ask whether nanoparticles could be harmful when inhaled or ingested. Clearly some nanoscale particles (such as asbestos fibres and some particulates produced from exhaust emissions) have deleterious effects connected with their size. But many nanoscaled materials have been in use for many years (and in the case of dispersions of nanoparticulates of natural origin, like milk, for much of human history) without ill effects.

The wide variety of nanoscaled materials and the variety of potential exposure routes suggests that it does not make sense to attempt to generalise about the putative toxicity or harmlessness of such materials as an entire class.

This question is currently of particular relevance to carbon nanotubes, as multi-walled nanotubes have some structural similarity to asbestos fibres.The number of published studies on nanotubes is small, and has not yet produced unequivocal evidence of toxicity.We should anticipate further studies on carbon nanotubes, and on other, newly introduced, nanoscaled materials, that should provide more definitive information to guide practices for working with and disposing of such materials before they enter bulk production.

Nanomaterial Science

Nanomaterials are not simply another step in the miniaturization of materials. They often require very different production approaches. There are several processes to create nanomaterials, classified as 'top-down' and 'bottom-up'. Although many nanomaterials are currently at the laboratory stage of manufacture, a few of them are being commercialised.

Below we outline some examples of nanomaterials and the range of nanoscience that is aimed at understanding their properties. As will be seen, the behaviour of some nanomaterials is well understood, whereas others present greater challenges.

One Dimensional Nano Materials

One-dimensional nanomaterials, such as thin films and engineered surfaces, have been developed and used for decades in fields such as electronic device manufacture, chemistry and engineering. In the silicon integrated-circuit industry, for example, many devices rely on thin films for their operation, and control of film thicknesses approaching the atomic level is routine. Monolayers (layers that are one atom or molecule deep) are also routinely made and used in chemistry. The formation and properties of these layers are reasonably well understood from the atomic level upwards, even in quite complex layers (such as lubricants). Advances are being made in the control of the composition and smoothness of surfaces, and the growth of films.

Engineered surfaces with tailored properties such as large surface area or specific reactivity are used routinely in a range of applications such as in fuel cells and catalysts. The large surface area provided by nanoparticles, together with their ability to self assemble on a support surface, could be of use in all of these applications.

Although they represent incremental developments, surfaces with enhanced properties should find applications throughout the chemicals and energy sectors. The benefits could surpass the obvious economic and resource savings achieved by higher activity and greater selectivity in reactors and separation processes, to enabling small-scale distributed processing (making chemicals as close as possible to the point of use). There is already a move in the chemical industry towards this. Another use could be the small-scale, on-site production of high value chemicals such as pharmaceuticals.

Two Dimensional Nanomaterials

Two dimensional nanomaterials such as tubes and wires have generated considerable interest among the scientific community in recent years. In particular, their novel electrical and mechanical properties are the subject of intense research.

(a) *Carbon Nanotubes*

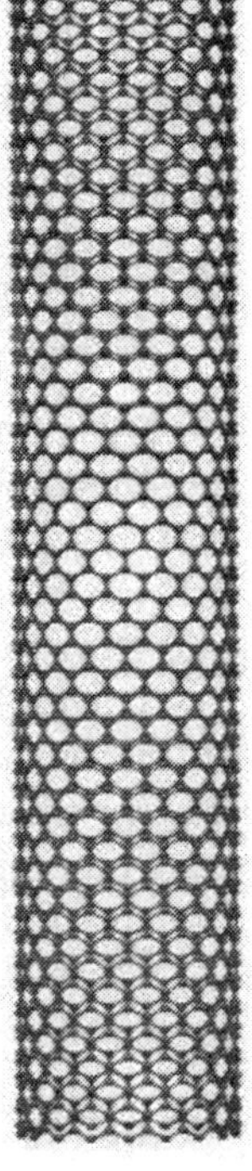

The Most Important Material in Nanotechnology Today: Carbon nanotubes (CNTs) were first observed by Sumio Iijima in 1991. CNTs are extended tubes of rolled graphene sheets. There are two types of CNT: single-walled (one tube) or multi-walled (several concentric tubes). Both of these are typically a few nanometres in diameter and several micrometres to centimetres long. CNTs have assumed an important role in the context of nanomaterials, because of their novel chemical and physical properties. They are mechanically very strong (their Young's modulus is over 1 terapascal, making CNTs as stiff as diamond), flexible (about their axis), and can conduct electricity extremely well (the helicity of thegraphene sheet determines whether the CNT is a semiconductor or metallic). All of these remarkable properties give CNTs a range of potential applications: for example, in reinforced composites, sensors, nanoelectronics and display devices.

Nanohorns

One of the SWNT (single walled carbon nanotube) types, with an irregular horn-like shape, which may be a critical component of a new generation of fuel cells. The main characteristic of the carbon nanohorns is that when many of the nanohorns group together an aggregate (a secondary particle) of about 100 nanometers is created. The advantage being, that when used as an electrode for a fuel cell, not only is the surface area extremely large, but also, it is easy for the gas and liquid to permeate to the inside. In addition, compared with normal nanotubes, because the nanohorns are easily prepared with high purity it is expected to become a low-cost raw material.

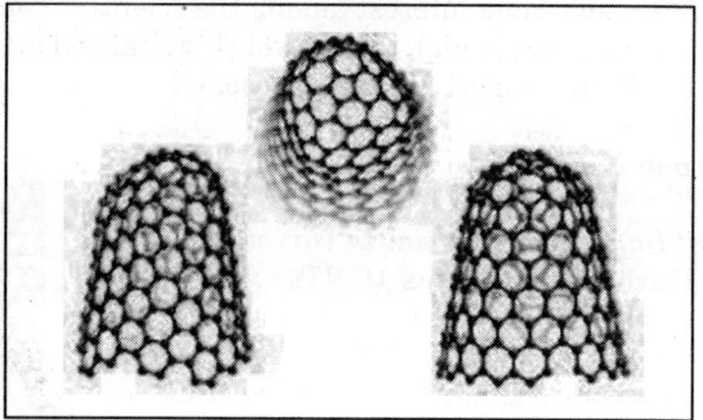

By using the minute and unique structure of the carbon nanohorn a tiny fuel cell for mobile terminals is developed.The fuel cell has about 10-times the energy capacity compared with a lithium battery, and if used for personal computers, in the future, a continued usage time of several days can be expected.

The fuel cells, which directly transform the chemical reaction energy between hydrogen and oxygen into electric energy, are seen as the energy source of the next generation. With their environmentally friendly and high efficiency characteristics the cells are being researched and developed as the future energy for the automobile and as energy generation for the home. The energy capacity is expected to become at least 10 times that of the present highest density lithium secondary battery, used in a wide range of applications. This new technological development is expected to solve the currently faced problems of dramatically reducing power consumption rates and is one big step towards achieving a practical fuel cell for mobile terminals.

The developed tiny fuel cell, classified as a polymer electrolyte fuel cell (PEFC), utilizes the carbon nanohorns as electrodes for catalyst support. It is observed that very fine platinum catalyst particles are dispersed on the surfaces of the carbon nanohorns. The size of the platinum particle is less than half of that supported on the ordinary activated carbon (acetylene black) by the same method. The size of the catalyst particle is one of the most important factors that determine the performance of the fuel cell, and it is considered that, the finer the size the better performance.

Although the reason a catalyst particle becomes fine is still not clear in the case of the carbon nanohorn, because of the unique shape of the aggregate it is thought that contact and grain growth of catalyst particles will be prevented. It is also expected that by further altering the form of the carbon nanohorn the dispersed state and the battery characteristic of the catalyst particle will improve. In addition, because a carbon nanohorn is produced by the laser ablation method, if a platinum catalyst is also simultaneously evaporated it is observed that a platinum particle will naturally adhere to the surface of a carbon nanohorn. If this method is used, the complicated catalyst supporting process through the conventional wet process can be omitted resulting in a large cost reduction.

Until now, since the discovery of the carbon nanotube, although it has been acknowledged as having a high possibility of being applied to semiconductors, flat-panel displays, lightweight and high-strength raw material and fuel cells etc., it had stopped at the fundamental research as a material stage. This development however, is the first step for the practical utilization of carbon nanotube and big steps towards the development and expansion of nano-technology.

(b) *Inorganic Nanotubes*

Inorganic nanotubes and inorganic fullerene-like materials based on layered compounds such as molybdenum disulphide were discovered shortly after CNTs. They have excellent tribological (lubricating) properties, resistance to shockwave impact, catalytic reactivity, and high capacity for hydrogen and lithium storage, which suggest a range of promising applications. Oxide-based nanotubes (such as titanium dioxide) are being explored for their applications in catalysis, photo-catalysis and energy storage.

(c) *Nanowires*

Nanowires are ultrafine wires or linear arrays of dots, formed by self-assembly. They can be made from a wide range of materials. Semiconductor nanowires made of silicon, gallium nitride and indium phosphide have demonstrated remarkable optical, electronic and magnetic characteristics (for example, silica nanowires can bend light around very tight corners). Nanowires have potential applications in high-density data storage, either as magnetic read heads or as patterned storage media, and electronic and opto-electronic nanodevices, for metallic interconnects of quantum devices and nanodevices. The preparation of these nanowires relies on sophisticated growth techniques, which include selfassembly processes, where atoms arrange themselves naturally on stepped surfaces, chemical vapour deposition (CVD) onto patterned substrates, electroplating or molecular beam epitaxy (MBE). The 'molecular beams' are typically from thermally evaporated elemental sources.

(d) *Biopolymers*

The variability and site recognition of biopolymers, such as DNA molecules, offer a wide range of opportunities for the self-organization of wire nanostructures into much more complex patterns. The DNA backbones may then, for example, be coated in metal. They also offer opportunities to link nano- and biotechnology in, for example, biocompatible sensors and small, simple motors. Such self-assembly of organic backbone nanostructures is often controlled by weak interactions, such as hydrogen bonds, hydrophobic, or van der Waals interactions (generally in aqueous environments) and hence requires quite different synthesis strategies to CNTs, for example. The combination of one-dimensional nanostructures consisting of biopolymers and inorganic compounds opens up a number of scientific and technological opportunities.

(e) *Nanoribbons*

As the name suggests, nanoribbons are solid objects (unlike nanotubes, which are hollow) with a nearuniform rectangular cross-section.

So far, nanoribbons have primarily been synthesized from the oxides of metals and semiconductors. In particular, SnO_2 and ZnO nanoribbons have been materials systems of great current interest because of potential applications as catalysts, in optoelectronic devices, and as chemical sensors for pollutant gas species and biomolecules. Although they grow to tens of microns long, the nanoribbons are remarkably single-crystalline and essentially free of dislocations. Thus they provide an ideal model for the systematic study of electrical, thermal, optical, and transport processes in one-dimensional semiconducting nanostructures, and their response to various external process conditions.

Recent experiments with SnO_2 nanoribbons indicate that these are highly effective in detecting even very small amounts of harmful gases like NO_2. Upon adsorption of these gases, the electrical conductance of the sample decreases by more than an order of magnitude. More interestingly, it is possible to get rid of the adsorbates by shining UV light, and the electrical conductance is completely restored to its original value. Such single crystalline sensing elements have several advantages over conventional thin-film oxide sensors: low operating temperatures, no ill-defined coarse grain boundaries, and high active surface-to-volume ratio.

Nano Materials in Three Dimensions

Fullerenes (Carbon 60)

Model C60. In the mid-1980s a new class of carbon material was discovered called carbon 60 (C60). Harry Kroto and Richard Smalley, the experimental chemists who discovered C60 named it "buckminsterfullerene", in recognition of the architect Buckminster Fuller, who was well-known for building geodesic domes, and the term fullerenes was then given to any closed carbon cage. C60 are spherical molecules about 1nm in diameter, comprising 60 carbon atoms arranged as 20 hexagons and 12 pentagons: the configuration of a football. In 1990, a technique to produce larger quantities of C60 was developed by resistively heating graphite rods in a helium atmosphere. Several applications are envisaged for fullerenes, such as miniature 'ball bearings' to lubricate surfaces, drug delivery vehicles and in electronic circuits.

Dendrimers

A dendrimer is a tree-like highly branched polymer molecule (Greek dendra = tree). Dendrimers are synthesized from monomers with new branches added in discrete steps ("generation") to form a tree-like architecture. A high level of synthetic control is achieved through step-wise reactions and purifications at each step to control the size, architecture, functionality and monodispersity. Several different kinds of dendrimers have been synthesized utilizing different monomers and some are commercially available. This picture shows a "3rd generation" polyamidoamine (PAMAM) dendrimer.

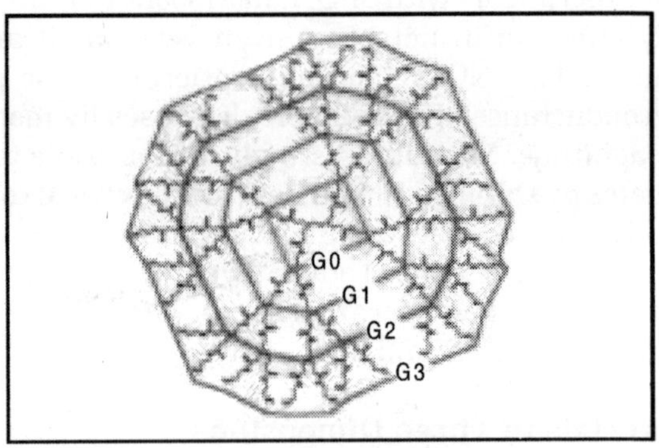

Dendrimers are used in conventional applications such as coatings and inks, but they also have a range of interesting properties which could lead to useful applications. For example, dendrimers can act as nanoscale carrier molecules and as such could be used in drug delivery. Environmental clean-up could be assisted by dendrimers as they can trap metal ions, which could then be filtered out of water with ultra-filtration techniques.

Dendrimers are of particular interest for cancer applications because of their defined and reproducible size, but more importantly, because it is easy to attach a variety of other molecules to the surface of a dendrimer. Such molecules could include tumor-targeting agents (including but not restricted to monoclonal antibodies), imaging contrast agents to pinpoint tumors, drug molecules for delivery to a tumor, and reporter molecules that might detect if an anticancer drug is working.

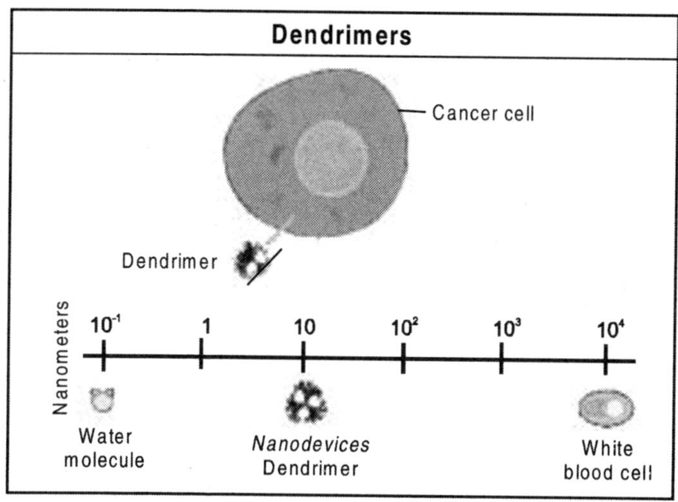

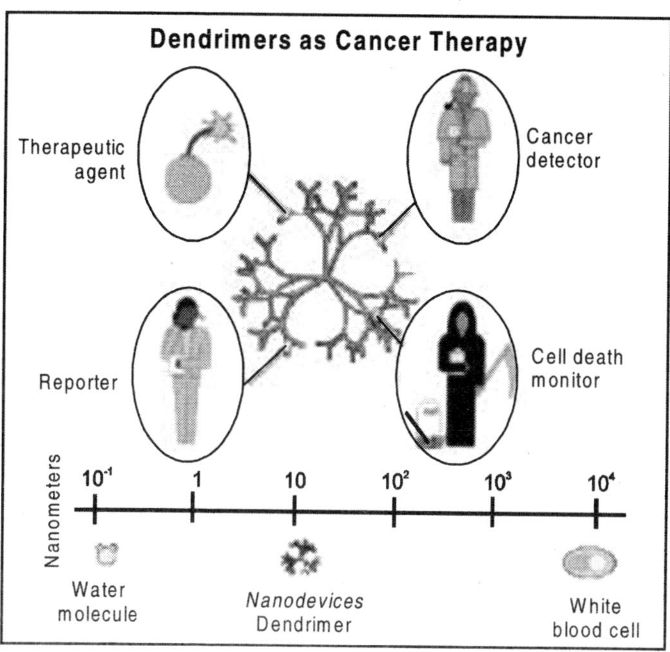

A single dendrimer can carry a molecule that recognizes cancer cells, a therapeutic agent to kill those cells, and a molecule that recognizes the signals of cell death. Researchers hope to manipulate dendrimers to release their contents only in the presence of certain trigger molecules associated with cancer. Following drug release, the dendrimers may also report back whether they are successfully killing their targets.

Nanotechnologies that will aid in cancer care are in various stages of discovery and development. Experts believe that quantum dots, nanopores, and other devices for detection and diagnosis may be available for clinical use in 5 to 15 years. Therapeutic agents are expected to be available within a similar time frame. Devices that integrate detection and therapy could be used clinically in about 15 or 20 years.

Quantum Dots

Ever since Faraday, in the 19th century, made a dispersion of nanosized particles of gold in water, chemists have been devising ways of creating such fine dispersions or colloids of a variety of materials.

The recent growth of interest in these materials has come about for three reasons. The chemistry has now been refined to an extent at which there is considerable control over both the size and the distribution of sizes of the particles. New physical techniques, including scanning probe microscopy, now permit the accurate characterisation of particle sizes. It has also been realised that the physical properties of such finely divided matter— particularly the electrical and optical properties—are strongly influenced by quantum effects. Nanoscale particles of semiconducting materials, such as cadmium selenide and gallium arsenide, are known as quantum dots—their size is such that quantum effects change the energy levels of their electrons.

This means that their optical and fluorescence spectra depend on their dimensions. In simpler terms, their colour changes with size. These have potential applications in new kinds of lasers and light emitting diodes.

Particles can be made to emit or absorb specific wavelengths (colours) of light, merely by controlling their size. Recently, quantum dots have found applications in composites, solar cells

(Gratzel cells) and fluorescent biological labels (for example to trace a biological molecule) which use both the small particle size and tuneable energy levels. Recent advances in chemistry have resulted in the preparation of monolayer-protected, high-quality, monodispersed, crystalline quantum dots as small as 2nm in diameter, which can be conveniently treated and processed as a typical chemical reagent.

They can also be used instead of dyes as markers for molecules in biological experiments. Similarly, chemical techniques are no w available to make rods of semiconducting material whose cross-section is of nanoscale dimensions.These quantum wires may be useful as components in molecular electronics.

To detect cancer, scientists can design quantum dots that bind to sequences of DNA that are associated with the disease. When the quantum dots are stimulated with light, they emit their unique bar codes, or labels, making the critical, cancer-associated DNA sequences visible.

The diversity of quantum dots will allow scientists to create many unique labels, which can identify numerous regions of DNA simultaneously. This will be important in the detection of cancer, which results from the accumulation of many different changes within a cell. Another advantage of quantum dots is that they can be used in the body, eliminating the need for biopsy.

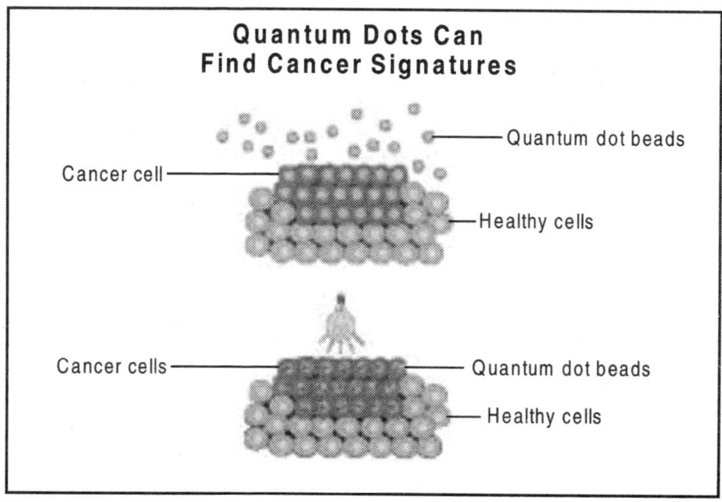

Major Nanomaterial Currently in Research and Development and Their Potential Applications

Material	Properties	Application	Time-Scale to Market Launch
Clusters of atoms			
Quantum wells	Ultra-thin layers-usually a few nanometres thick—of semiconductor material (the well) grown between barrier material by modern crystal growth technologies. The barrier materials trap electrons in the ultra-thin layers, thus producing a number of useful properties. These properties have led, for example, to the development of highly efficient laser devices.	CD players have made use of quantum well lasers for several years. More recent developments promise to make these nanodevices commonplace in low-cost telecommunications and optics.	Current 5 years
Quantam dots	Fluorescent nanoparticles that are invisible until 'lit up' by ultraviolet light. They can be made to exhibit a range of colours, depending on their composition	Telecommunications, optics.	7-8 years
Polymers	Organic-based materials that emit light when an electric current is applied to them and vice versa	Computing, energy conversion	?

(Contd.)

(Contd.)

Material	Properties	Application	Time-Scale to Market Launch
Grains that are less than 100 nm in size			
Nanocapsules	Buckminsterfullerenes are the most well known example. Discovered in 1985, these C60 particles are 1nm in width	Many applications envisaged e.g. nano-particulate dry lubricant for engineering	Current 2 years
Catalytic nanoparticles	In the range of 1-10 nm, such materials were in existence long before it was realised that they belonged to the realms of nano-technology. However, recent developments are enabling a given mass of catalyst to present more surface area for reaction, hence improving its performance. Following this, such catalytic nano-particles can often be regenerated for further use.	Wide range of applications, including materials, fuel and food production, health and agriculture	Current-?
Fibres that are less than 100 nm in diameter			
Carbon nanotubes	Two types of nanotube exist: the single-wall carbon nanotubes, the so-called 'Buckytubes', and multilayer carbon nanotubes. Both consist of graphitic carbon and typically have an internal diameter of 5 nm & an external diameter of 10 nm. Described as the 'most important material in nanotechnology today', it has been calculated that nanotube-based material has the potential to become 50-100 times stronger than steel at one sixth of the weight.	Many applications are envisaged: space and aircraft manufacture, automobiles and construction. Multi-layered carbon nanotubes are already available in practical commercial quantities. Buckytubes some way off large-scale commercial production.	Current 5 years

(Contd.)

(*Contd.*)

Material	Properties	Application	Time-Scale to Market Launch
Films that are less than 100 nm in thickness			
Self-assembling monolayers (SAMs)	Organic or inorganic substances spontaneously form a layer one molecule thick on a surface. Additional layers can be added, leading to laminates where each layer is just one molecule in depth.	A wide range of applications, based on properties ranging from being chemically active to being wear resistant	2-5 years
Nanoparticulate coatings	Coating technology is now being strongly influenced by nano-technology. E.g. metallic stainless steel coatings sprayed using nano-crystalline powders have been shown to possess increased hardness when compared with conventional coatings.	Sensors, reaction beds, liquid crystal manufacturing, molecular wires, lubrication and protective layers, anticorrosion coatings, tougher and harder cutting tools.	5-15 years

(*Contd.*)

(Contd.)

Material	Properties	Application	Time-Scale to Market Launch
Nanostructured materials			
Nano-composites	Composites are combinations of metals, ceramics, polymers and biological materials that allow multi-functional behaviour. When materials are introduced that exist at the nanolevel, nano-composites are formed and the material's properties—e.g. hardness, transparency, porosity—are altered.	A number of applications, particularly where purity and electrical conductivity characteristics are important, such as in microelectronics. Commercial exploitation of these materials is currently small, the most ubiquitous of these being carbon black, which finds widespread industrial application, particularly in vehicle tyres.	Current 2 years
Textiles	Incorporation of nano-particles and capsules into clothing leading to increased lightness and durability, and 'smart' fabrics (that change their physical properties according to the wearer's clothing)	Military, lifestyle.	3-5 years

8. TWO TYPES OF CONSTRUCTIONS

(A) Top-Down

From top (larger) to bottom (smaller). Mechanisms and structures are miniaturized to a nanometric scale. This has been the most frequent application of nanotechnology up to this point, in particular in the domain of electronics where miniaturization is preponderant.

The top-down approach is analogous to making a stone statue. You take a bulk piece of material and modify it, by carving or cutting in the case of stone, until you have made the shape you want. The process involves material westage and is limited by the resolution of the tools you can use, restricting the small sizes of the structures made by these techniques.

Examples of this kind of approach include the various types of lithographic techniques (such as photo-, ion beam-, electron- or X-rays-lithograph) cutting, etching and grinding.

Lithography is a selective process that allows the patterning of a desired design onto the starting material. A mask is used to protect the areas of material that need to be maintained, while the rest of the material is exposed to a beam of UV, ions, electrons or X-rays. Etching and/or deposition on the surface is then performed, leaving the desired shape.

These techniques have been used to great effect to produce the miniaturisation of electronic components, such as computer chips, MEMS (micro-electromechanical systems) in consumer products such as computer hard drives, and CD and DVD players.

Lithography has fuelled the continued miniaturisation of these components so that Moore's law has continued to be applicable, namely that the number of transistors per integrated circuit doubles every couple of years.

We will study the techniques in detail in next Chapter.

Quantum well lasers and high-quality optical mirrors are also fabricated using top-down techniques.

(B) Bottom-Up

From bottom (smaller) to top (larger). We begin with a nanometric structure such as a molecule, and through a process of assembly or self-assembly we create a mechanism larger than that with which we began. This approach, considered by some to be the one and only "true" nanotechnology, should allow an extremely precise control of matter. It is in this way that we will be able to free ourselves from the limits of miniaturization, notably in the domain of electronics.

There is less wastage with this technique, and strong covalent bonds will hold the constituent parts together. However, this technique is limited in how big the structures can be made.

A good example of this kind of approach is found in nature; all cells use enzymes to produce DNA by taking the component molecules and binding them together to make the final structure. Chemical synthesis, self-assembly, and molecular fabrication are all examples of bottom-up techniques.

Self-assembly occurs when little manipulation occurs, but components spontaneously come together to form ordered nanoscale structures, for instance in the formation of nanotubes and some monolayers.

Molecular fabrication is the only one of these techniques where molecules or atoms are manipulated into position one by one. Consequently this is a laborious process which has yet to address the problems of scaling up the technique for manufacture. It has been speculated that this process could be scaled up using self-replicating machines that would allow parallel production, leading to further speculation about the replication process running out of control. Recently however, Eric Drexler, the scientist who first proposed these scenarios, has retracted his suggestion of out-of-control replication leading to 'gray goo'.

Cosmetics, fuel additives, and experimental atomic and molecular devices are all examples of structures made using bottom-up techniques.

The limits of feature size and quality of these two approaches have started to converge in recent years, leading to new hybrid techniques of manufacture.

9. SPINTRONICS

Spintronics is emerging to be a new form of nanotechnologies, which utilizes not only the charge but also spin degree of freedom of electrons. Spin-dependent tunneling transport is one of the many kinds of physical phenomena involving spintronics, which has already found industrial applications. The use of tunneling magnetoresistive reading heads has helped to maintain a fast growth of areal density, which is one of the key advantages of hard disk drives as compared to solid-state memories. This review is focused on the first commercial tunneling magnetoresistive heads in the industry at an areal density of 80~100 Gbit/in^2 for both laptop and desktop Seagate hard disk drive products using longitudinal media. The first generation tunneling magnetoresistive products utilized a bottom stack of tunnel junctions and an abutted hard bias design. The output signal amplitude of these heads was 3 times larger than that of comparable giant magnetoresistive devices, resulting in a 0.6 decade bit error rate gain over the latter. This has enabled high component and drive yields. Due to the improved thermal dissipation of vertical geometry, the tunneling magnetoresistive head runs cooler with a better lifetime performance, and has demonstrated similar electrical-static-discharge robustness as the giant magnetoresistive devices. It has also demonstrated equivalent or better process and wafer yields compared to the latter. The tunneling magnetoresistive heads are proven to be a mature and capable reader technology. Using the same head design in conjunction with perpendicular recording media, an areal density of 274 Gbit/in^2 has been demonstrated, and advanced tunneling magnetoresistive heads can reach 311 Gbit/in^2. Today, the tunneling magnetoresistive heads have become a mainstream technology for the hard disk industry and will still be a technology of choice for future hard disk products. We will learn more on Spintronics in chapter 6.

10. MOLECULAR NANOTECHNOLOGY

The more hyped aspects of nanotechnology have generally revolved around MNT.

Proponents of this approach suggest that environmentally clean, inexpensive, and efficient manufacturing of structures, devices, and 'smart' products, based on the flexible control of architectures and processes at an atomic or molecular scale of precision, may be feasible in the near future (i.e. 10–20 years from the present). The ambitious goal is to produce complex products on demand using simple raw materials, such as by inserting the basic chemical elements in a molecular assembly factory to yield a common household appliance . These visions have attracted a great deal of public interest, and impressive demonstrations have been made of microscopic devices. For example, in August 2001 scientists from Osaka University built the smallest micromechanical system ever, a spring whose arm is only 0.3 μm wide However, although almost qualifying as a nanodevice, the question of whether it is possible to attain extreme capability and, if so, how to develop the field, is a point of contention in both scientific and policy circles.

Inspite of the above controversies, it remains clear that bottom-up technologies, while having the potential to be immensely important in the longer term, are not likely in the near future However, some products benefiting from research into molecular manufacturing may be developed in the near term. As initial nanomachining, novel chemistry and protein engineering (or other biotechnologies) are refined, initial products will likely focus on those that substitute for existing high-cost, lowerefficiency products. Likely candidates for these technologies include a wide variety of sensor applications, tailored biomedical products (including diagnostics and therapeutics), extremely capable computing and storage products, and unique, tailored materials (i.e. smart materials using nanoscale sensors, actuators, and perhaps controller elements) for aerospace or similar high-capability needs . Predictions of when bottom-up processes will begin to become available on a widespread basis vary across the literature. In general, the hyped aspects of the industry are operating around a 20-year time-scale, with estimates for economically viable self-assembly techniques tending to convene around 2015. However, to reach a fully mature nanotechnology society—where it is possible to manipulate objects on all scales from atom to macroscopic—is expected to take at least 35 years. This is partly due to the economic advantages of competing technologies. For example, with regard to advanced computing, Anton *et al.*, (2001) state that: '*the odds-on favourite for the next 15 years remains traditional digital electronic computers based on semiconductor technology. Given the virtual certainty of continued progress in this area, it is hard to imagine a scenario in which...quantumswitch-based computing, molecular computers, or something else could offer a significant performance advantage at a competitive price.'* The major technical obstacles to development in other areas of MNT— namely molecular manufacturing, general assembly and nanobots—are expanded below.

(i) Molecular Manufacturing

To realise molecular manufacturing, a number of technical accomplishments are necessary). First, suitable molecular building blocks must be found. These building blocks must be physically durable, chemically stable, easily manipulated, and (to a certain extent) functionally versatile. The second major area for development is in the ability to assemble complex structures based on a particular design. A number of researchers have been working on different approaches to this issue. One uses atomic force or molecular microscopes with very small nanoprobes to move atoms or molecules around with the aid of physical or chemical forces. An alternative approach uses lasers to place molecules in a desired location.

Chemical assembly techniques are also being addressed, including an approach to building structures one molecular layer at a time. A third major area for development within molecular manufacturing is systems design and engineering. Extremely complex molecular systems at the macroscale will require substantial subsystem design, overall system design, and systems integration, much like complex manufactured systems of the present day. Although the design issues are likely to be largely separable at a subsystems level, the amount of computation required for design and validation is likely to be quite substantial. Performing checks on engineering constraints, such as defect tolerance, physical integrity, and chemical stability, will be required as well .

The long term goal of molecular manufacturing is to build exactly what we want at low cost. Many if not most of the things that we'll want to build are complex (like a molecular computer), and seem difficult if not impossible to synthesize with currently available methods. Adding programmed positional control to the existing methods used in synthesis should let us make a truly broad range of macroscopic molecular structures. To add this kind of positional control, however, requires that we design and build what amount to very small robotic manipulators. If we are to make anything of any significant size with this approach, we'll need mole quantities of these manipulators. Fortunately, any truly general purpose manufacturing device should be able to manufacture another general purpose manufacturing device, which lets us build large numbers of such devices at low cost. This general approach, used by trees for a very long time, should let us develop a low cost general purpose molecular manufacturing technology.

(ii) Nanobots and Other Nanoscale Devices

This area can be accredited with receiving the most severe hype, where headline grabbing predictions include curing cancer, eliminating infections, enhancing our intelligence, and even making us immortal. In fact, according to Saxl, it will take 25 years at least before tiny machines circulate in the bloodstream cleaning out fat deposits from our arteries. Indeed, although the implications of such revolutionary technologies are awesome, developments that appear achievable in the short and medium term are not particularly dramatic. Perhaps the most advanced work in this area concerns MIT's Bioinstrumentation Laboratory where an autonomous miniature robot, dubbed the 'NanoWalker', is being designed . Measuring approximately 25 mm^2, the name NanoWalker stems from its ability to take thousands of steps per second in the nanometre range. The ultimate goal of this type of robotic machine, generically referred to as an assembler, is the construction of materials an atom or molecule at a time by precisely placing reactive groups. This is called 'positional assembly'

(iii) General Assembly

The General Assembler is considered to be the 'Holy Grail' of nanotechnology and represents the ultimate utility of automanipulating nanobots. In general, such an assembling device is regarded as extremely distant (e.g. more than 25 years).

11. NANOSTRUCTURED MATERIALS BY SELF-ASSEMBLY

Self-Replication

The act of objects making copies of themselves—similar to biological reproduction except that self-replicating objects should make exact copies of themselves

An example from nature is a virus; a computer virus is a self-replicating computer program; the polymerase chain reaction (PCR) is an example of in vitro self-replication. While self-replication makes for great science fiction, it is relatively irrelevant to real nanotechnology and it's importance in the next several decades is expected to be minimal. The reason for this is that the vast majority of objects that self-replication might be able to fabricate can be fabricated far more easily without self-replication.

Take PCR for example. This is a method by which if one has even a single DNA molecule, it is possible to make a copy from each molecule in vitro , resulting in a sufficient amount of DNA to observe macroscopically. Furthermore, the tool used to cary out this reaction (a thermocycler) might be considered to be an assembler, since it carries out a series of specific chemical reactions and molecular assembly steps in order to produce the DNA output. However, since the purpose of this reaction is primarily to simply read the DNA sequence, nanotechnology will likely provide a more direct (and cheaper) method of reading the DNA, making self-replication a useful trick for the time being, but in general not the most efficient approach. Perhaps one day, it might be feasible to create a new life form from the molecular level, but seeing as we still know very little about the life already in existance it seems that we have much to learn first. Alias: Reproduction.

Molecules such as soap, which have two or more sections with a distinctly different chemical character (amphiphiles), can form complex nanostructured phases by self-assembly.The key feature of the resulting materials is that they are hierarchical.A number of molecules come together to make a structure, such as a sphere or a rod, with dimensions in the nanometre range, and then these units themselves assemble in a regular way. Such structures – known as surfactant mesophases – are widely exploited in personal products, such as shampoos and hair gels, as well as in cosmetics and pharmaceutical preparations. Soap has always been nanostructured, and soap-boilers have over the centuries crafted knowledge about how the compositions of soaps can be adjusted to give different physical properties. More recently the relationship between their nanoscale structure and properties has been elucidated, permitting a much greater degree of rational design. The nanoscale structures that are formed by soap-like molecules are intrinsically soft, but they can be used as templates for the synthesis of hard materials which will then have a precisely controlled nanoscale porosity. Such materials, which have a very high ratio of surface to bulk, are attractive candidates for new, efficient catalyst materials, and offer the potential to improve all sorts of chemical engineering processes. Polymeric molecules may also have an amphiphilic character which allow them to self assemble into complex morphologies whose basic units have dimensions on the nanometre scale. Some of these materials are already commercialised, for example as thermoplastic elastomers—materials with the elasticity and resilience of rubber, but which, unlike rubber, can be melted and moulded repeatedly.

In the future such materials could be used in optoelectronics as photonic crystals and in biomedical science as scaffolds for artificial skin and organs.

Perhaps the most specific and finely engineered examples of selfassembly are to be found in nature, in the folding of proteins and the base-pairing mechanism of DNA. Protein folding has some superficial resemblances to the type of self-assembly that occurs in block copolymers with a water soluble block and a hydrophobic block. In water, such a molecule will fold up in a way that keeps the hydrophobic part of the molecule in the centre, sheltered from the water by an outer layer formed from the hydrophilic block. A protein molecule also has hydrophobic units and in water it folds in such a way as to keep the hydrophobic parts out of the way of the water. But rather than folding into any one of a large number of roughly similar arrangements, as a block copolymer would do, a protein folds into a completely defined shape.This is the shape that is precisely optimised for catalysing chemical reactions or for being a component in a molecular machine. At the moment the physics of this process is just beginning to become understood. Creating synthetic molecular machines by a similar route is an attractive target but will require major steps forward in both physics and chemistry. The self-assembly mechanism that underlies the operation of DNA is much simpler to understand. Attached to the backbone of a single strand of the DNA molecule is a sequence of 'bases', structures can be quite complicated and rich in their topology; for example in a combination of oil, water, and a soap-like molecule it is possible to arrive at a structure in which the oil and water are each of which is chosen from one of four (conventionally represented by the letters T, C, A and G). These four bases make up two complementary pairs; A binds strongly to T, and C to G.

Thus for every distinct strand of DNA, defined by a sequence of bases (CTCAGGACT, say), there is a complementary strand (in this case GAGTCCTGA); these two complementary strands will associate very strongly in the famous double helix structure.

Because synthetic DNA can now be made with arbitrary sequences of bases, one can imagine making assemblies of DNA molecules that are programmed to come together in specific shapes. Quite complex shapes have been made in this way; these shapes can be used as scaffolds on which other materials can be deposited, opening the way to the creation by self-assembly of intricate, three-dimensional nanoscale structures.

12. NANOPHOTONICS

In this emerging area, Photonic band gap (PBG) materials (also called as photonic crystals) are an interesting class of materials which are constituted of a periodically varying dielectric constant (refractive index) and thus show a band gap for light transmission. The range of wavelengths that are absent in transmission will be present in reflection, when the constituent materials are non-absorbing in that range. The band gap region is decided by the modified Bragg's law, which relates the high reflection wavelength to the periodicity and the effective index. The periodicity can be in one-, two or three dimensions. The standard methods employed for this are based on semiconductor technology, such as lithography. Self-assembly is a simpler and cheaper alternative where spherical colloids, monodispersed in an appropriate solvent, assemble under thermodynamic equilibrium into 3D-ordered crystals. The main drawback of self-assembly is the number of unwanted defects in the crystal during the process of ordering.

The structural characterization with a resolution of 1 mm will show the close-packed arrangement of spheres and the optical characterization will show the peak reflectance wavelength. It is established that the self-assembled samples of colloidal spheres usually arrange themselves in an fcc structure with the plane parallel to the substrate. To achieve a photonic stop band in the visible, one requires the period to be less than a micron. The photonic crystals have many interesting applications in optoelectronics and optical communications such as filters, low-threshold lasers, waveguides and photonic crystal fibers. When a defect is introduced in a photonic crystal, it forms a resonant cavity so that light is localized at the defect. If the "defect" contains an active species that emits light under excitation, then the cavity effect is capable of modifying the spontaneous and stimulated emission characteristics. High Q cavities are required for the fabrication of low-threshold lasers from such active species.

13. ELECTRONICS AND OPTOELECTRONICS

Our second major theme is in the area of electronics and optoelectronics, which underlie the information technology and communications industries. Modern consumer electronics is already approaching the nanoscale, and the drive to continue the spectacular advances in capability of electronics that have occurred over the last 25 years is a major driver for research programmes in nanotechnology. Continuous incremental improvement is taking place in industrial laboratories, in the technology for making integrated circuits for central processing units and memory chips, and this will continue.

Meanwhile, in academic laboratories, the groundwork is being laid for new technologies that may take over from the current ones somewhere between ten and 20 years hence.

Semiconductor Opto-Electronics

Modern communications technologies—high capacity telephone networks and the internet—have been built on a combination of digital electronics and the use of light to carry data in optical fibres. The interaction between electronics and light and the conversion of information-bearing signals between the two media is the realm of optoelectronics. Here much effort has been devoted to controlling the structure of semiconductors on the nanoscale to create new light-emitting diodes and lasers. This is already a commercial technology, and most people will be aware of the much wider use of light emitting diodes as light sources following the development of reliable blue emitters. These together with existing green and red devices permit the generation of white light far more efficiently than with conventional incandescent or fluorescent light bulbs.

Memory and Data Storage

Computer data storage has progressed over the last 50 years from the highly macroscopic—punched cards—towards the microscopic, with magnetic tapes and disks. In these materials information is stored as a pattern of magnetised regions. In a hard disk, information is stored on tracks separated by a few microns. Along the track a single bit of information might be stored on a length of track of less than 100 nm, giving a net storage density of one Gbit per square centimetre. This can be compared with a storage density of about 360 kbit per square centimetre in a floppy disk, a factor of about 3000.

This remarkable feat of miniaturisation has been made possible by a combination of extreme precision microengineering and the development of very sensitive read heads, which depend on the phenomenon of giant magnetoresistance. This very sensitive response of the electrical properties of a material to an applied magnetic field can be obtained in composite materials consisting of nanometre thick multilayers of metals with different magnetic properties.

Given the very rapid progress made in conventional magnetic storage, where a mature solution (by the fast moving standards of information technology) is being refined under the pressure of very large markets, one might wonder whether there is any point in looking for a radically different solution. Nonetheless, should physical limits prevent the continuation of recent trends in miniaturisation, some alternative approaches are beginning to take shape. One possible approach is to use self-assembled structures, made using block copolymers, as templates for arrays of magnetic particles in the 10 nm range. This could potentially lead to increases in data density of another factor of 1000, but there are a number of potential difficulties to be overcome related to the lack of perfection in the long ranged order of these structures.

The other familiar storage device is the optical disk—the compact disk or DVD. Here information is stored as a pattern of craters in a flat polymer surface, which are read by a laser beam.The ultimate limit on the density of craters, and thus on the density of information that can be packed on the surface of such a device, is set by the wavelength of laser light. This limit has already been reached in DVDs. Some further incremental gains can be obtained by using blue light, because of its smaller wavelength. But to take this conceptually straightforward approach to information storage, in which we use physical marks on a surface, to smaller sizes we would need radically different ways both of writing and of reading the information. One such approach would be to use an atomic force microscope (AFM). IBM has developed a system whereby modified AFM heads 'write' marks in a plastic surface by heating, to melt small pits. The pits are read by a parallel array of more than 10,000 heads.The ultimate in data density would be achieved if a unit of information could be stored in a single molecule.

New Methods for Data Input and Output

Although these words are being processed by computers of a complexity unimaginable 50 years ago, the method by which the words are transferred from the authors' brains to the machine—a typewriter keyboard—and from the machine to the reader—a piece of paper—represent very old technology indeed. Although the virtues of keyboards and paper are very great, it seems hard to imagine that different and more powerful methods for getting information into and out of computers will not be developed.

For some of these methods—speech recognition, for example— the problems are essentially ones of software, so NST (Nano scale technology) will only contribute indirectly in as much as they will be helped by the general increase in computer power. On the input side, inputs from the physical world are obtained from a variety of sensing technologies—for example of temperature, chemical composition and pressure—and latest researchs will lead to smaller and more sensitive sensors, capable, for example, of detecting biochemicals in the blood without the need to remove it from the body. Developments in the human interface should greatly benefit the disabled, though much technology will be developed by the defence industry in an attempt to make the interface between a soldier or airman and a weapons system even smoother. Such interfaces could rely on the direct detection of electrical signals in the brain, or a physical connection between the nervous system and semiconductor logic.

On the output side, there have been substantial improvements in display technologies, with cathode ray tubes now essentially obsolete, replaced by liquid crystal displays, plasma displays, and field emission displays.

The point to make here is that developments in NST are leading to displays that are cheaper, larger, brighter, and more efficient than current ones. A more radical display technology would form an image directly on the retina. An interesting combination of the old and the new is provided by so-called electronic inks; these combine the crispness and contrast of printing onto paper with the switchability of an emissive or liquid crystal display. Particles of electronic ink can be switched from black to white by the application of an electrical field.

Usually the output from a computer is now an image or a piece of text, but since computers are used so frequently to design artefacts, it would be useful to provide an output that was the three-dimensional artefact itself. Various technologies for rapid prototyping achieve just this goal. This can be done by repeated ink-jet printing using, instead of ink, a material that solidifies after printing, to build up a three-dimensional image made from plastic.

Alternatively, a container of liquid monomer can be scanned by a laser beam, whose light initiates the polymerisation of the monomer to form a solid plastic object.

14. PLASTIC ELECTRONICS

Many readers will have noticed that as time goes by computers seem to get very much faster, but not significantly cheaper. This reflects the fact that efforts in the information technology industry has been devoted to increasing performance almost at any cost (speaking here of the capital cost of semiconductor plants).

One area moving against this trend is the field of plastic electronics, in which semiconducting polymers are used as the active materials to make logic circuits and display devices. Semiconducting polymers have significantly worse electronic properties than conventional semiconductors like silicon or gallium arsenide. But they are very cheap to make. Rather than using expensive lithography to pattern them to form circuits, very cheap processes such as ink-jet printing or soft lithography can be used. They can also be made into devices that are flexible. Currently there are three major areas of research.

Firstly, polymer light emitting diodes have been commercialised already; the potential here is for large area, flexible display devices such as roll-up computer or TV screens to be made cheaply. Entirely new products, such as clothing with an electronic display, can also be envisaged.

Secondly, field effect transistors are the basic element of logic and memory circuits, so very low cost printed logic circuits would be possible, with potential applications in packaging.

One can imagine a radio frequency ID device being incorporated in packaging or in an artefact that would identify itself and announce its presence. This would have a major impact on the way supply chains are managed in manufacturing industry, retail and distribution.

Finally, being actively researched, but currently furthest away from commercialisation, are photovoltaic devices, such as solar cells. In fact solar cells made from semiconducting polymers form just one class of novel solar cell architectures made from unconventional nanostructured semiconductors; Grätzel cells, made from nanosized titanium dioxide particles (a cheap and widely available substance which is the major pigment in white paint) sensitised by organic dyes are another example, and variants are being experimented with which use C60 particles. Once again, the efficiencies and lifetimes of these unconventional solar cells currently compare poorly with conventional materials based on inorganic semiconductors.

However, if significant improvements can be made the processing technologies that would be available to make very large areas of material cheaply could transform the economics of renewable energies, bringing down the cost of energy generated by solar cells towards the cost of non-renewable sources such as gas and oil.

15. MOLECULAR ELECTRONICS

One way to overcome the limits that lithography suffers on feature size would be to use conducting molecules as wires and as the elements of active components such as transistors, diodes, and switches.These could be carbon nanotubes or conducting polymers of the kind developed for plastic electronics. Although some fascinating preliminary work has been done to showing that transistors can be made from carbon nanotubes, significant difficulties remain. Although it is now clear that a semiconducting device based on a single molecule can be made, it is not at all clear how such devices could be wired together to make logic systems. One possibility would be to use the base pairing mechanism of DNA to design connections that would form by self-assembly.

These devices will work quite differently from larger scale transistors, because new physics comes into play at these very small length scales. But when understood properly, this new physics will give rise to new opportunities. An example is the Coulomb blockade effect, which arises from the fact that in a very small system the injection of a single electron can effectively change the electrical properties of that system.This effect can be used to make memories that depend on the presence or absence of a single electron.

The idea of a molecule whose state can be switched between two configurations has also been suggested as a unit of memory. The most developed examples of such molecules are the rotaxanes, in which a molecule ring is threaded on a spindle, along which it can be shuttled from one position to another. Again, although this idea has been shown to work in principle, the practical problem is in addressing the individual molecules.

Nanotubes and semiconductor nanowires are likely to be important components of a new molecular based computer architecture. Field effect transistors based on nanotubes have already been reported, and progress has been made towards integration using the flow of fluids on the nanoscale to align the nanowires.

16. BIOMEDICAL SCIENCE

If the biggest economic driving force for nanotechnology now comes from information technology, the area with the most resonance with the general public probably comes from the possibility of applying new technology to medicine. This area combines the very incremental with the highly futuristic. Some of the sharpest ethical and social dilemmas arising from nanoscale science and technology will have their origin in the success in this programme.

Because so many major applications for nanotechnology are likely to come from medicine, there is going to be a general need to understand the interaction between artificial nanodevices and living systems. This will be an important barrier to the use of nanotechnology for medicine; the body has extremely well developed mechanisms for recognising foreign bodies and neutralising them, and these mechanisms will in turn need to be overcome if our nanodevices are to do their job.

Drug Delivery

A high proportion of drugs that are administered today are delivered to the body in ways that would have been familiar to physicians one hundred years ago; orally, by injection, or by inhalation. Nanotechnology promises to yield much more sophisticated and precisely targeted ways of delivering drug molecules to the relevant part of the body. This may be necessary simply as a consequence of the widening search for new drug molecules. Highly insoluble compounds may be considered; these may be made usable by being prepared in the form of nanoscale particles. Classes of compounds that cannot survive passage through the stomach, like polypeptides, may become more widely used. For these, means of delivery other than frequent injection would be desirable. Other potentially useful drugs, particularly anti-cancer agents, may have very unpleasant and dangerous side effects, which could be minimised if the agents could be delivered selectively to their destination without affecting other parts of the body. Finally, gene therapy combines an extremely fragile and delicate molecule—DNA—with the need to target destination cells rather precisely. In many of these cases what is needed is a way of wrapping up a molecule, either to protect it from a hostile environment, or to protect the environment from the unwanted side-effects of the molecule.

One approach to this task which exploits NST is to use the soft nanostructures known as liposomes or vesicles.These are synthetic enclosures made from self-assembled bilayers of amphiphilic (soap-like) molecules, rather like very crude synthetic cells.

The basic liposome that one would obtain by ultrasonic treatment of a phospholipid solution is relatively fragile.They are used in the formulation of high-value cosmetics.Various developments on the basic theme provide structures which are considerably more useful for pharmaceuticals. If water-soluble polymer chains are incorporated in the structure they can greatly reduce the interactions that the liposome has with the environment. In the medical context this means that they will circulate much longer in the bloodstream before the body's methods for dealing with foreign bodies destroys them.These have been termed 'stealth liposomes' and this drug delivery technology is already commercialised. Simple liposomes are rather fragile objects, but they can be made more robust by chemically linking their components to form two dimensional solid sheets, or by using polymeric amphiphiles (block copolymers).The ideal is to make a delivery vehicle that would be robust until triggered to release its contents, and various strategies are emerging to achieve this.

The ultimate robust and selective delivery vehicle is a virus, whose entire purpose is to introduce its own genetic material into a target cell. Currently gene therapy relies on the use of viruses to introduce the necessary genetic material into a cell.This process is not without its dangers, so a synthetic analogue would be very desirable.

Tissue Engineering

The obvious limitations of organ transplants are that there are not enough organ donors, and that problems of rejection by the host immune system are very severe.The solution would be to grow new organs from cells provided by the host.

This is not easy, because although all the cells in a body have the same genetic blueprint—that is to say, the same DNA—cells are social organisms and they need to be persuaded to grow into skin or bone or a liver or whatever organ is required.

Tissue engineering attempts to get round this problem by providing a scaffold for the cells, which defines the structure that is required to produce the organ in question. The scaffold will be made from a biocompatible or natural polymer which has been patterned on the microscale or nanoscale, and which may have its surface treated in a way that makes the cells respond in the desired way.

Although the goal of making organs like hearts and livers is still quite far away, the tissue engineering of skin for grafts is well advanced and in use in clinical practise.

The Laboratory-on-a-Chip

Classical chemistry and biochemistry carry out their operations on rather large scales; if these scales could be miniaturised this could increase the sensitivity of chemical analysis and the ease with which it could be automated. A laboratory-on-a-chip would combine very small scale manipulation of chemicals with sensitive detection and direct interfacing to a computer for automatic control and analysis of the results.

The manipulation of liquids on small length scales involves some interesting new problems. Very small channels and reaction vessels can be created by etching patterns on the surface of glass or silicon. At these small scales, however, the movement of fluids is dominated by viscosity, making it difficult to get the fluids eitherto flow or to mix. Progress in microfluidics (see next section), as this field is known, will involve developing ways of overcoming these difficulties; electrophoresis, for example, can be used to move fluids using applied electric fields, and gels which respond to their environment by swelling or shrinking can be used to provide switchable valves. An example of the way in which the detection of very small quantities of a wide variety of target molecules can already be achieved is provided by the DNA microarray. In this technology, thousands of spots of different sections of DNA (each corresponding to a single gene) are printed in an array onto a surface. The aim of the analysis is to find out which of these genes are expressed in a group of cells under study; to do this the messenger RNA from the cells (or mRNA, the intermediary in the process of making the protein corresponding to each gene) is extracted.The complementary DNA sections corresponding to these mRNA molecules are made, fluorescently labelled and washed over the microarray.The presence of mRNA corresponding to each expressed gene is detected as a fluorescent signal at the corresponding point of the microarray.

This technology illustrates how self-assembly combined with fluorescent detection of surface associated molecules can lead to the simultaneous detection of very many analytes with very high sensitivity. It could in principle be scaled down to considerably smaller dimensions than are currently used, given the extremely high sensitivity (down to single molecules) with which fluorescent signals can be detected. The automation and scaling down in size of chemical processes has been tremendously powerful, and achievements such as the sequencing of the human genome have relied on this approach. Nonetheless, sequencing DNA is still resource and labour intensive. A seductive vision in nanotechnology is the idea of taking a single DNA molecule and physically reading off the sequence. In fact this vision is not entirely far-fetched; one can certainly manipulate single DNA molecules using techniques such as atomic force microscopy and laser traps. Having isolated a single DNA molecule, one approach to sequencing it would simply be to read along it with an atomic force microscope. Another would be to thread the molecule through a pore in a membrane, detecting small changes in, for example, the ionic conductivity of the pore as the molecule was pulled through.Whatever the details, given our rapidly growing capacity to handle single molecules of DNA, it seems likely that some physical (and therefore fast and cheap) method of reading the DNA sequences will be developed in the relatively near future.This would greatly speed up the progress of molecular biology, but would also open up new challenges and opportunities in medicine. If the complete genome of any individual were readily available then screening for all kinds of diseases with a genetic component would become very straight forward indeed.

Microfluidics

A multidisciplinary field comprising physics, chemistry, engineering and biotechnology that studies the behavior of fluids at volumes thousands of times smaller than a common droplet.

Microfluidic components form the basis of so-called "lab-on-a-chip" devices that can process microliter and nanoliter volumes and conduct highly sensitive analytical measurements. The fabrications techniques used to construct microfluidic devices are relatively inexpensive and are amenable both to highly elaborate, multiplexed devices and also to mass production. In a manner similar to that for microelectronics, microfluidic technologies enable the fabrication of highly integrated devices for performing several different functions on the same substrate chip. Microfluidics is a critical component in gene chip and protein chip development efforts.

17. THE QUANTUM CONCEPT OF TUNNELING

When we observe nature at the nanoscale, certain phenomena become apparent. Electrons simultaneously act as particles and as waves—a key concept of quantum mechanics. This behavior helps explain the "tunneling phenomenon."

Classical physics tells us that for electrons in solids to go from point A to point B, they must have a continuous conductor along which to travel. A classic textbook example is electricity traveling through wire. If the electron current is to move from one wire to another, the wires must touch in order for the current to complete its path.

Quantum mechanics challenges this concept. Physical contact of the two wires is not necessary, as long as the gap between them is approximately one nanometer. When the gap is this small, the electrons can "extend" over to the other side.

So, why do scientists call this "tunneling?" The electrons are essentially tunneling through a barrier. Though a barrier made of space may seem counterintuitive to us, the little electrons much prefer metal, so a gap of space would normally be a roadblock for them.

Imaging

What does "tunneling" have to do with taking images? It is important to remember that the images produced by the STM are not like photographs. The STM is not like a camera recording an atom's light emission on film—it is a tool that "senses" rather than sees. Because of tunneling currents, the STM does not have to touch the sample in order to make a mapping.

The tip is positioned at a fixed, nanoscale distance away from the sample surface. A small voltage is applied between the tip and the substrate to "encourage" the tunneling current. When the tip is close to the substrate, the tunneling current increases and more electrons "leap" over. When the tip is farther away from the substrate, the tunneling current decreases and fewer electrons "make the leap." Since the researcher fixes the current, the topography of the substrate is responsible for variations in the tip-to-surface relationship, and those variations are recorded and translated into colorful, computerized profiles.

With these profiles, researchers can examine the quality of atomically layered surfaces, or they can study the properties of specific molecules

18. NANOCRYSTALS

Nanocrystals are particularly attractive as building blocks for larger structures because it's possible—even easy—to prepare nanocrystals that are highly perfect."

Nanocrystals are aggregates of anywhere from a few hundred to tens of thousands of atoms that combine into a crystalline form of matter known as a "cluster." Typically around ten nanometers in diameter, nanocrystals are larger than molecules but smaller than bulk solids and therefore frequently exhibit physical and chemical properties somewhere in between. Given that a nanocrystal is virtually all surface and no interior, its properties can vary considerably as the crystal grows in size.

By precisely controlling a nanocrystal's size and surface, its properties can be tuned, we can tune the bandgap, how it conducts charge, we can change what crystal structure it resides in, we can even change its melting temperature.

Growing flawless nanocrystals is relatively easy because their length of scale is so small there's simply not enough time during the growth process to introduce defects. This same tiny length of scale, however, makes controlling the size and surface of nanocrystals a tremendous challenge. During the past decade, Scientist have been growing nanocrystals out of semiconductor powders and exploring various ways of altering growth conditions as a means of meeting this challenge.

Scientist get first big breakthroughs when they discovered that spherical nanocrystals made from a core of cadmium selenide inside a shell of cadmium sulfide could, depending upon their size, be made to emit multiple colors of light. This opened the door to a number of potential applications, including the use of these spherical core-shell nanocrystals as highly effective fluorescent labels for the study of biological materials. In fluorescent labeling, markers, usually antibodies that attach themselves to specific proteins, are tagged with dye molecules that fluoresce or emit a specific color of light when stimulated by photons, usually from a confocal microscope.

Recently scientist grow semiconductor nanocrystals into two-dimensional rods that paves the way for a slew of new potential applications, and also proves that controlling crystal growth is the key to controlling shape as well as size.

More recently, scientist have learned to manipulate the conditions and rate of crystal growth to the point where they have obtained semiconductor nanocrystals in the shape of tear drops, arrowheads, and even four-armed tetrapods. While these exotic shapes have no immediate application, they expand the possible things that might be built from nanocrystal blocks in the future. For example, when the tetrapod nanocrystals are dropped onto a surface they always land on three arms with the fourth arm pointing straight up. This should be a handy feature for the wiring of nanosized electronic devices

In the familiar "twisted ladder" image of DNA, two strands of phosphate and ribose sugar molecules are joined by "rungs" made up of a connecting pair of nitrogenous compounds called "bases." There are four types of bases-adenine (A), cytosine (C), guanine (G), and thymine (T)-and A always pairs with T, and G always connects with C. Scientist capitalize on this highly specific architectural program by using "linker" molecules to attach segments of single-stranded DNA up to 100 bases in length (about 33 nanometers) to crystals of gold measuring 5 to 10 nanometers across. When these nanocrystal/DNA "conjugates" are mixed with other segments of single-stranded DNA containing base sequences that complement the sequences of the DNA in the conjugates, the complementary bases recognize one another and pair off to form double-stranded DNA. In this manner, DNA serves as a template for creating nanocrystal molecules.

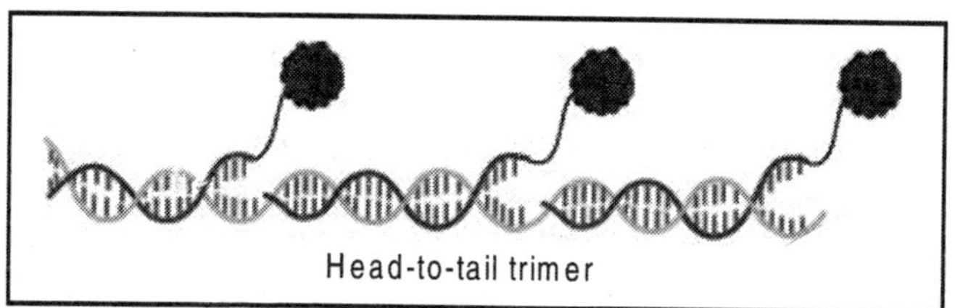

DNA is being used as a template for assembling nanocrystals into macromolecules. Here nanocrystals of gold attached to a single strand of DNA are brought together into a desired alignment as the DNA's double-helix is formed.

The shift in particle mobility (in the electrified gel) due to the DNA attachment can be used to produce nanocrystal/DNA conjugates with a well-defined number of DNA strands.

Proof that the information contained within DNA can be harnessed for the spatial patterning of semiconductor nanocrystals holds great promise for their use as nanotechnology building blocks. Although the assemblies so far have been relatively simple, the indication is that nanocrystal/DNA conjugates could be used to make structures and devices corresponding in size and complexity to the semiconductor circuits produced today using lithography.

Nanocrystals might be used to make super-strong and long-lasting metal parts. The crystals also might be added to plastics and other metals to make new types of composite structures for everything from cars to electronics.

Thin Films

Thin films are atomically engineered layers of a wide variety of materials including metals, insulators and semiconductors. The major applications of thin films are in modification of the surface properties of solids. Individual films may be electrically conductive or non-conducting, hard or soft, thermally conducting or insulating, optically transparent, or opaque. A thin film coating can transform the electrical, mechanical and/or optical properties of a solid base material in a cost-effective way. Some common examples are scratch-resistant coatings for spectacles, anti-reflection coatings for lenses, transparent conducting coatings for flat-panel displays, and low-friction coatings for bearings. Hard coatings can significantly enhance the lifetime of cutting, drilling, and forming tools. Oxygen and moisture barrier films are in widespread use in the packaging of foodstuffs, contributing to the long shelf life of many convenience foods. Thin film coatings also have unique properties that may be exploited in the polarization, reflection, transmission and absorption of light. Complex coatings can be used to provide eye-protection from lasers without significant reduction in overall transmission and other high-performance films are in use for the multiplexing of telecommunication laser signals. Other inherent properties of thin films are used in microelectronics, magnetic recording and optical recording media.

What Nanodevices can do in Medical Field

Nanodevices are Small Enough to Enter Cells

Most animal cells are 10,000 to 20,000 nanometers in diameter. This means that nanoscale devices (less than 100 nanometers) can enter cells and the organelles inside them to interact with DNA and proteins. Tools developed through nanotechnology may be able to detect disease in a very small amount of cells or tissue. They may also be able to enter and monitor cells within a living body.

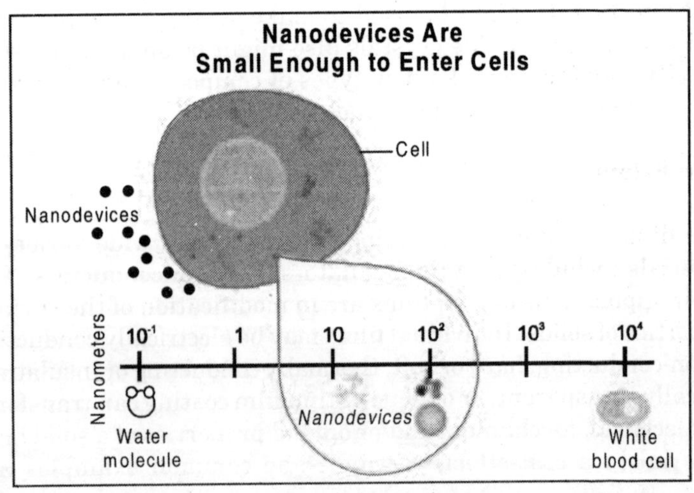

Nanodevices can Improve Cancer Detection and Diagnosis

Detection of cancer at early stages is a critical step in improving cancer treatment. Currently, detection and diagnosis of cancer usually depend on changes in cells and tissues that are detected by a doctor's physical touch or imaging expertise. Instead, scientists would like to make it possible to detect the earliest molecular changes, long before a physical exam or imaging technology is effective. To do this, they need a new set of tools.

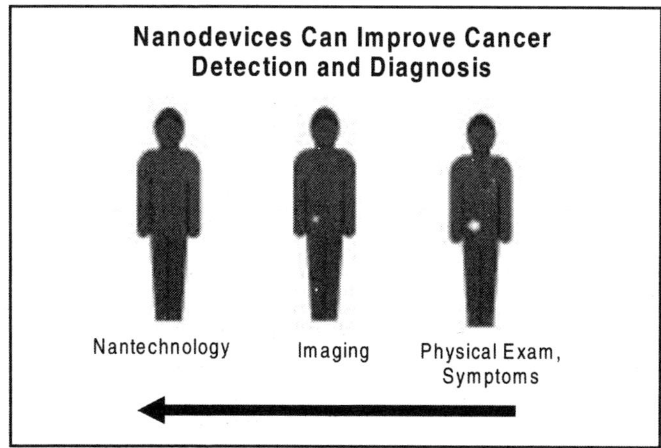

Nanodevices can Improve Sensitivity

In order to successfully detect cancer at its earliest stages, scientists must be able to detect molecular changes even when they occur only in a small percentage of cells. This means the necessary tools must be extremely sensitive. The potential for nanostructures to enter and analyze single cells suggests they could meet this need.

Nanodevices can Preserve Patients' Samples

Many nanotechnology tools will make it possible for clinicians to run tests without physically altering the cells or tissue they take from a patient. This is important because the samples clinicians use to screen for cancer are often in limited supply. It is also important because it can capture and preserve cells in their active

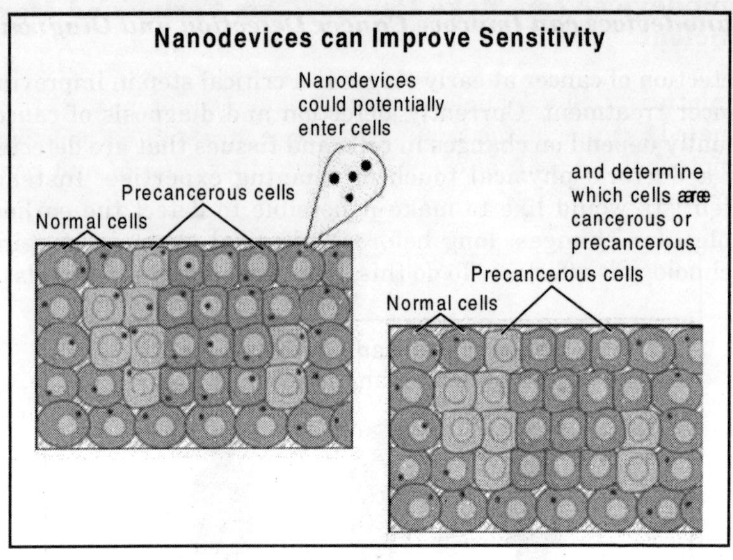

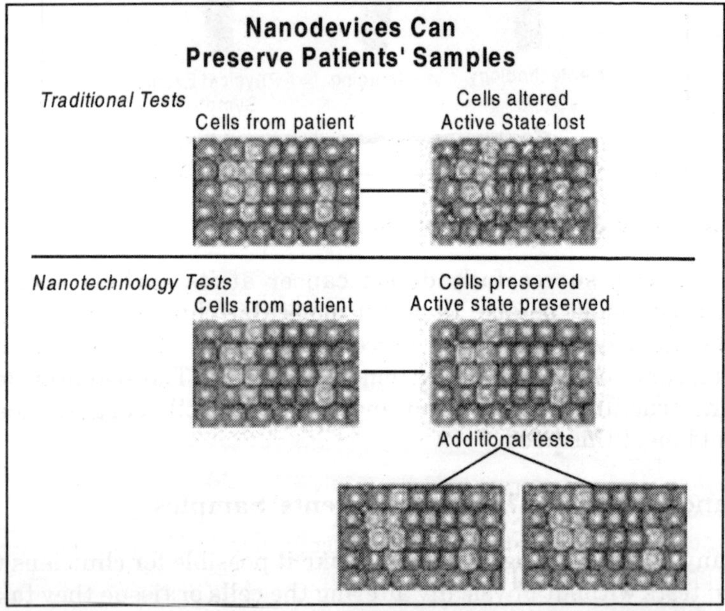

state. Scientists would like to perform tests without altering cells, so the cells can be used again if further tests are needed.

Nanodevices can Make Cancer Tests Faster and More Efficient

Miniaturization will allow the tools for many different tests to be situated together on the same small device. Researchers hope that nanotechnology will allow them to run many diagnostic tests simultaneously.

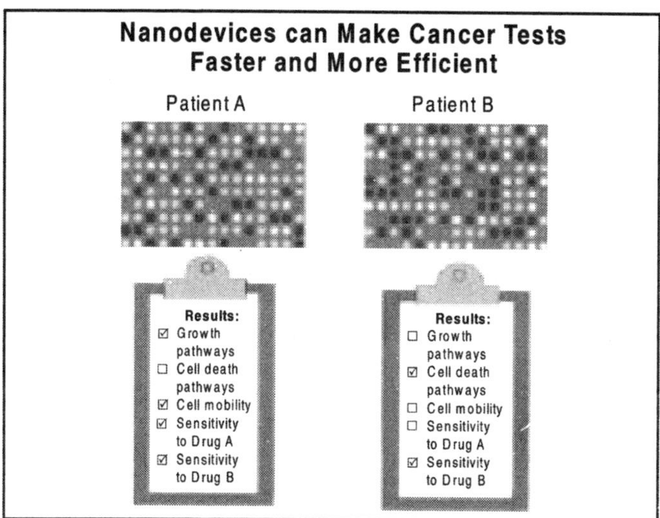

Cantilevers can Make Cancer Tests Faster and More Efficient

One nanodevice that can improve cancer detection and diagnosis is the cantilever. These tiny levers, which are anchored at one end, can be engineered to bind to molecules that represent some of the changes associated with cancer. They may bind to altered DNA sequences or proteins that are present in certain types of cancer. When these molecules bind to the cantilevers, surface tension changes, causing the cantilevers to bend. By monitoring the bending of the cantilevers, scientists can tell whether molecules are present. Scientists hope this property will prove effective when cancer-associated molecules are present—even in very low concentrations—making cantilevers a potential tool for detecting cancer in its early stages.

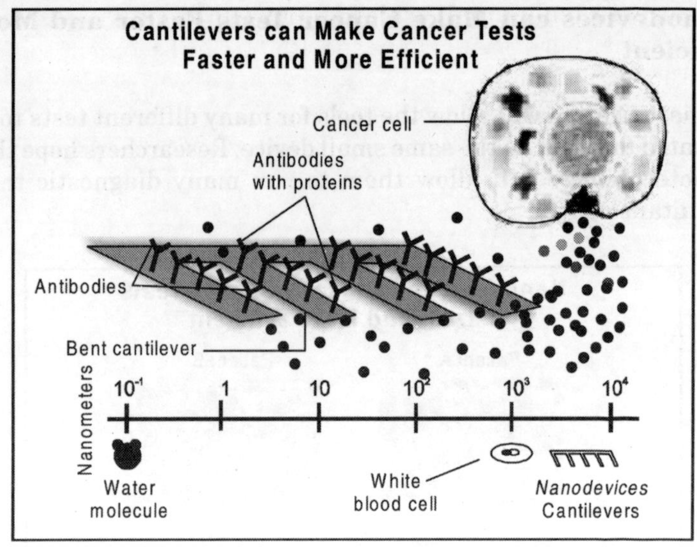

Nanopores

Another interesting nanodevice is the nanopore. Essentially bitty tiny holes. Nanoscopic pores found in purpose-built filters, sensors, or diffraction gratings to make them function better. As activated carbon, they may also be used as an alternative fuel storage medium, due to their massive internal surface area. Improved methods of reading the genetic code will help researchers detect errors in genes that may contribute to cancer. Scientists believe nanopores, tiny holes that allow DNA to pass through one strand at a time, will make DNA sequencing more efficient. As DNA passes through a nanopore, scientists can monitor the shape and electrical properties of each base, or letter, on the strand. Because these properties are unique for each of the four bases that make up the genetic code, scientists can use the passage of DNA through a nanopore to decipher the encoded information, including errors in the code known to be associated with cancer.

In biology, they are "complex protein assemblies that span cell membranes and allow ionic transport across the otherwise impermeable lipid bilayer. Nanopores are important because while some pores help maintain cell homeostasis, others disrupt cell function." "A nanopore can be a protein channel in a lipid bilayer or an extremely small isolated 'hole' in a thin, solid-state membrane" such that "DNA and RNA, can be registered and characterized singly.

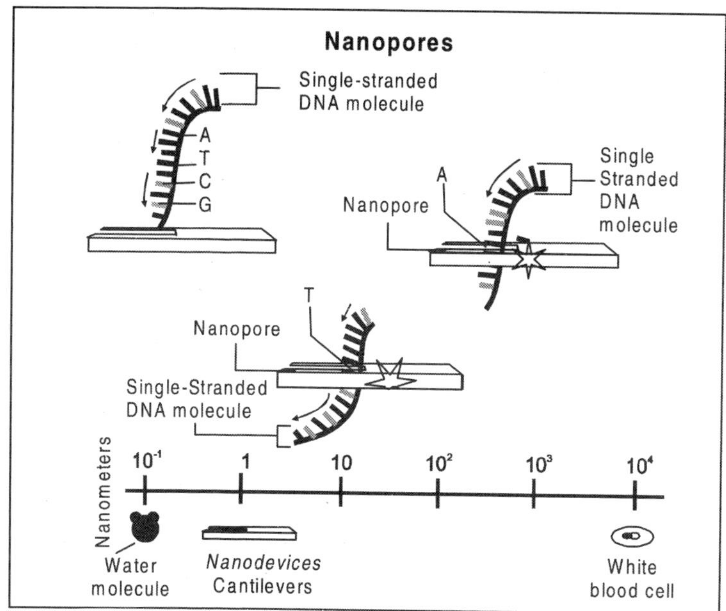

Nanotubes—Marking Mutations

Another nanodevice that will help identify DNA changes associated with cancer is the nanotube. Nanotubes are carbon rods about half the diameter of a molecule of DNA that not only can detect the presence of altered genes, but they may help researchers pinpoint the exact location of those changes.

To prepare DNA for nanotube analysis, scientists must attach a bulky molecule to regions of the DNA that are associated with cancer. They can design tags that seek out specific mutations in the DNA and bind to them.

Once the mutation has been tagged, researchers use a nanotube tip resembling the needle on a record player to trace the physical shape of DNA and pinpoint the mutated regions. The nanotube creates a map showing the shape of the DNA molecule, including the tags identifying important mutations. Since the

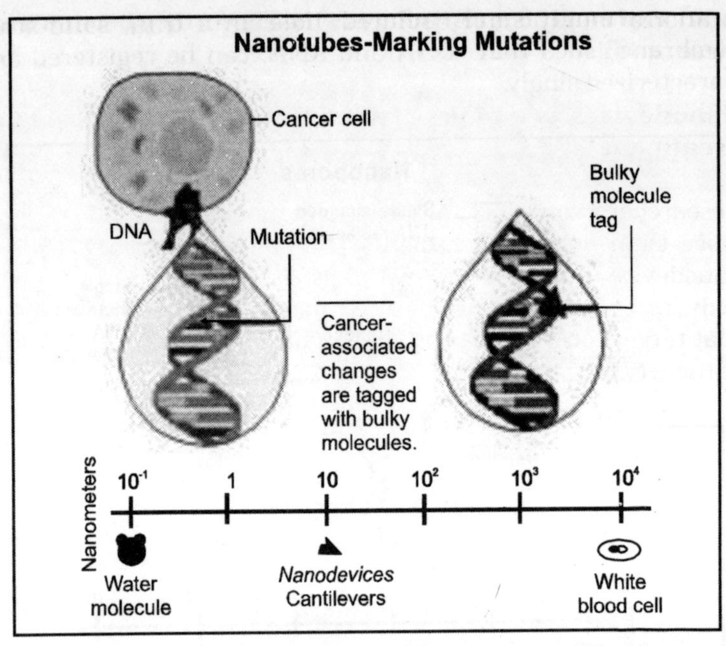
Nanotubes-Marking Mutations
Cancer cell
DNA
Mutation
Bulky
molecule
tag
Cancer-
associated
changes
are tagged
with bulky
molecules.
Nanometers
10⁻¹
1
10
10²
10³
10⁴
Water
molecule
Nanodevices
Cantilevers
White
blood cell

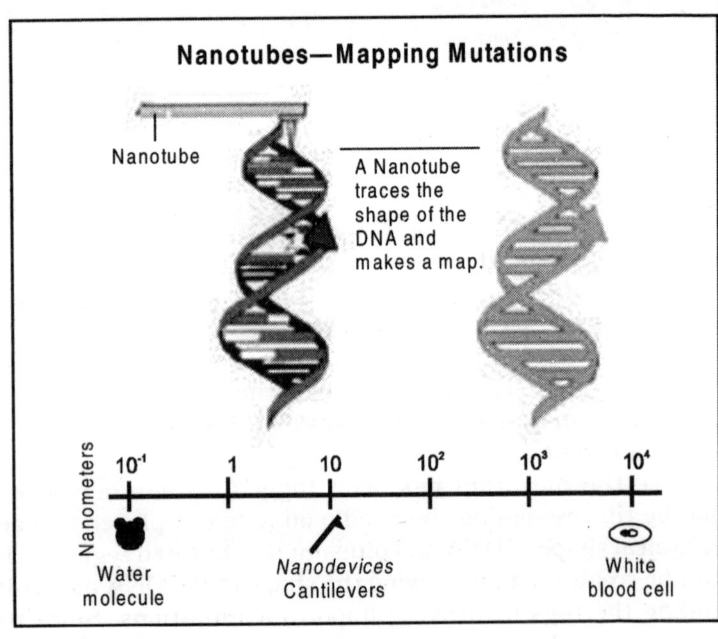
Nanotubes—Mapping Mutations
Nanotube
A Nanotube
traces the
shape of the
DNA and
makes a map.
Nanometers
10⁻¹
1
10
10²
10³
10⁴
Water
molecule
Nanodevices
Cantilevers
White
blood cell

location of mutations can influence the effects they have on a cell, these techniques will be important in predicting disease.

Nanodevices as a Link Between Detection, Diagnosis, and Treatment

Researchers aim eventually to create nanodevices that do much more than deliver treatment. The goal is to create a single nanodevice that will do many things: assist in imaging inside the body, recognize precancerous or cancerous cells, release a drug that targets only those cells, and report back on the effectiveness of the treatment.

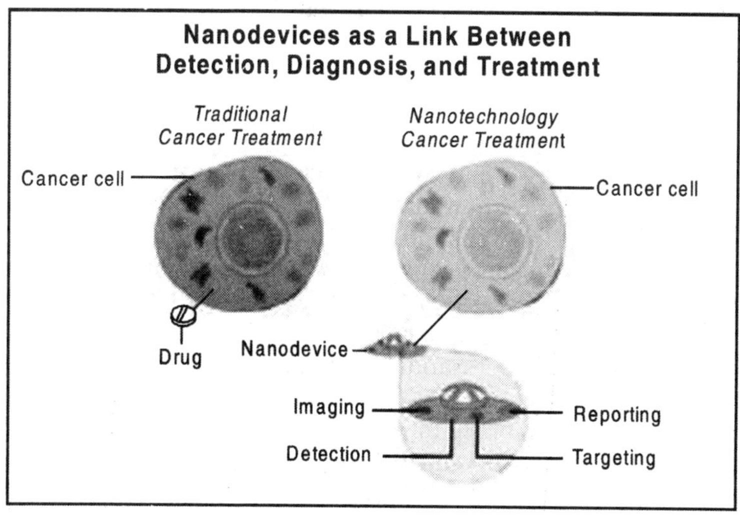

TOOLS TO MEASURE AND MAKE NANOSTRUCTURES

The essence of nanotechnology is the ability to work at the atomic, molecular and macromolecular levels in order to create materials, devices and systems with fundamentally new properties and functions. Building blocks are atoms and molecules, or their assemblies such as nanoparticles, nanolayers, nanowires and nanotubes. The relative arrangement of the elementary blocks of matter into their assemblies leads to new properties and functions even for the same chemical composition. For example, the arrangement of the carbon atoms at nanoscale is the only difference between soft graphite, hard diamond, and conducting nanotubes. Machines with complex functions on the scale of a virus or a human cell are envisioned.

Establishing understanding and manufacturing methods at the building blocks of matter is a historical opportunity in human development. The ability to rearrange matter on a nanoscale is potentially a very economical way to obtain functionality, with the promise of becoming the highest-added-value manufacturing approach. The matter can be rearranged at this scale by using weak interactions, such as electrostatic dipole, hydrogen bonds, van der Walls forces, hydrophobic/hydrophilic interactions, complimentary DNA hybridization, fluidic assembly, and other assembling and patterning approaches. Guided selfassembling is an example where the arrangement of molecules is made under control by an external magnetic field, electric field, flow field, templating or other means. Manipulation of matter with atomic/molecular precision by weak interactions requires relatively low energy dissipation, significantly lower that changes at the subatomic level or changes at larger scales for obtaining the same property or function. It may become the ultimate manufacturing approach once one would achieve fundamental understanding of phenomena and processes at that scale.

The main scientific drivers are discovery of new phenomena at nanoscale, methods of measurements and modeling of large number of nano-objects, understanding the connection between nanostructure and function, manipulation with atomic and molecular precision, assembling and connecting at nanoscale, understanding modern biology and the synergism with information technology.

Physical, chemical, biological, materials and engineering sciences have arrived to nanoscale about the same time. Engineering plays an important role because when we refer to nanotechnology we speak about 'systems' at nanoscale, where the treatment of simultaneous phenomena in multibody assemblies would require integration of disciplinary methods of investigation and an engineering system approach. The manipulation of a large system of molecules is equally challenging to a thermodynamics engineer researcher as it is to a single-electron physics researcher. They need to work together. Engineering needs to redefine its domain of relevance to effectively take this role in conjunction with other disciplines. Several reasons for an increased role of engineering are:

Nanotechnology deals with systems at nanoscale, which are hierarchically integrated in architectures at larger scales. Typically, the number of components is in the range of hundreds or thousands or tens of thousands. Such components are defined by their collective behavior.

Multiple phenomena act simultaneous. Nanotechnology requires the integration of the methods of investigation from various disciplines in order to understand macroscopic phenomena, define transport coefficients, optimize processes and design products. Various methods need to be considered at different length scales. For instance, multiscale modeling of dynamic fracture would require finite element simulation of continuum elasticity, then atomistic simulation of Newton's equation and thereafter electronic simulation of Schrodinger's equation.

Nanotechnology implies the ability to manipulate the matter under control at the nanoscale and integrate manufacturing along scales. Main challenges are creation of tailored structures at the nanoscale, and combination of the bottom-up and top-down approaches to generate nanostructured devices and systems. Further challenges are integration of living and non-living structures, replication and eventually self-replication methods at nanoscale, and development of new concepts that would allow economic scale-up for industrial production. Development of tools and processes to measure, calibrate and manufacture.

The engineering community needs to redefine the role of engineering from analysis, design and manufacturing mainly at the macro- and micro-scales towards the 'nanoscale engineering'

Six increasingly interconnected megatrends in science and engineering are perceived as dominating the scene for the next decades:

    (i)   Information and computing
    (ii)   Nanoscale science and engineering
    (iii)   Biology and bio-environmental approaches
    (iv)   Medical sciences and eventually enhancing human physical capabilities
    (v)   Cognitive sciences concerned with exploring and enhancing intellectual abilities
    (vi)   Collective behavior and system approach to study nature, technology and society.

Such advancements evolve in coherence, with multiple areas of confluence and with temporary divergences. For example, information technology helps to simulate and visualize the nanoworld, and nanoscale tools help measurement and manipulation of DNA. Nanoscale science and engineering is expected to grow in close synergism with the digital revolution and modern biology. It is the most exploratory and a condition for the development of the other two in the next 10-15 years. Melding of human development with science and engineering development is also notable. Fundamental discoveries and revolutionary innovations at nanoscale create a tension between the society's quest for more control over nature in the future, and society's strong desire for stability and predictability in the present. The development of nanoscale science and engineering is considered in this broader perspective.

TOOLS AND FABRICATION

It is a simple statement of fact that in order to make things you must first have the fabrication tools available. Therefore, many of the nanomaterials are coevolving with a number of enabling technologies and techniques. These tools provide the instrumentation needed to examine and characterize devices and effects during the R&D phase, the manufacturing techniques that will allow the large-scale economic production of nanotechnology products, and the necessary support for quality control Because of the essential nature of this category, its influence is far greater than is reflected in the size of the economic sectors producing these products.

Tools and Techniques

That can be Used to Measure and Make Nanostructures

- Microscopy
- Metrology
- Simulation
- Crystallography
- Interferometry
- Chemical Synthesis
- Plasma & other regimens
- Lithography

MICROSCOPY

It may inclued:

- Acoustic/Ultrasonic
- Fluorescent/UV
- Laser/Confocal
- Polarizing
- Portable Field
- Scanning Electron Microscope (SEM)
- Scanning Probe/Atomic Force (SPM/AFM)
- Transmission Electron Microscope (TEM)
- Scanning Near-Field Optical Microscope (SNOM)

The resolution of an optical microscope is about a third of a wavelength in diameter, which is about 200 nm.

- Acoustic, 30 micrometers at 50Mhz, 200 Mhz to maybe 20 nm The resolution of a SEM is about 10 nanometers (nm).
- The resolution of a TEM is about 0.2 nanometers (nm). This is the typical separation between two atoms in a solid.
- The optical resolution limit for SNOM is governed by the light intensity passing through the aperture. A practical limit is usually found with aperture diameters between 80 nm and 200 nm, but in ideal cases even down to **< 20** nm.

Some of the best values for AFM imaging are 3.0 nm. Sub-nanometer is possible

Later in the chapter we will study scanning probe/atomic force in detail

METROLOGY

Fundamental to commercial nanotechnology is repeatability, and fundamental to repeatability is measurement.

Nanometrology, then, allows the perfection of the texture at the nanometre and subnanometre level to be examined and controlled. This is essential if highly specialised applications of nanotechnology are to operate correctly, for example X-ray optical components and mirrors used in laser technologies:

- Metrology involves
  - ➤ Critical Dimension Measurement
  - ➤ Film Thickness Testers
  - ➤ Resistivity/Electromagnetic Testers
  - ➤ Stress Measurement
  - ➤ Wafer Inspection Tools
  - ➤ Quantum measurements

Partial list of commercially available techniques employed in particle measurement:

- Acoustic Attenuation Spectroscopy
- Laser Doppler Velocimetry (LDV)
- Aerosol Mass Spectroscopy (Aerosol MS)
- Light Microscopy or Optical Imaging
- Condensation Nucleus Counter (CNC)
- Microelectrophoresis
- Differential Mobility Analysis (DMA)
- Scanning Electron Microscopy (SEM)
- Electrical Zone Sensing (Coulter Counting)
- Sedimentation (Gravitational & Sentrifugal)
- Electroacoustic Spectroscopy
- Sieving
- Electrokinetic Sonic Amplitude
- Tapered Element Oscillating Microbalance (TEOM)
- Gas Adsorption Surface Area Analysis (e.g. BET)
- Transmission Electron Microscopy (TEM)
- X-ray Diffraction (XRD)

Dynamic Light Scattering (DLS) or Photon Correlation Spectroscopy (PCS) or Quasielastic Light Scattering (QELS).

SIMULATION

- Simulation includes
  - molecular modeling
  - kinetic modeling
  - quantum effect modeling
  - semiconductor effects

Later in the chapter we will study molecular modelling in detail

CRYSTALLOGRAPHY

- Crystallography
  - x-ray

INTERFEROMETRY

- Interferometry
  - optical

> x-ray
> quantum

CHEMICAL SYNTHESIS

- Chemical Synthesis
  > organic
  > biological
  > genomic

PLASMA

Techniques

- Plasma & other regimens
  > coatings
  > materials fabrication
  > surface treatments

LITHOGRAPHY

- Lithography
  > manufacturing
  > prototyping/testing

Later in the chapter we will study lithography in detail.

Let us first study fabrication of most important material of nanotechnology carbon nanotubes.

CARBON NANO TUBE(CNT)—FABRICATION

Carbon Arc or Arc Discharge

A vacuum chamber is pumped down and back filled with some buffer gas, typically neon or Ar to 500 torr. A graphite cathode and anode are placed in close proximity to each other. The anode may be filled with metal catalyst particles if growth of single wall nanotubes is required. A voltage is placed across the electrodes, (20-40V). The anode is vaporized while the cathode evaporates. Carbon nanotubes form on the cathode in the sheath region.

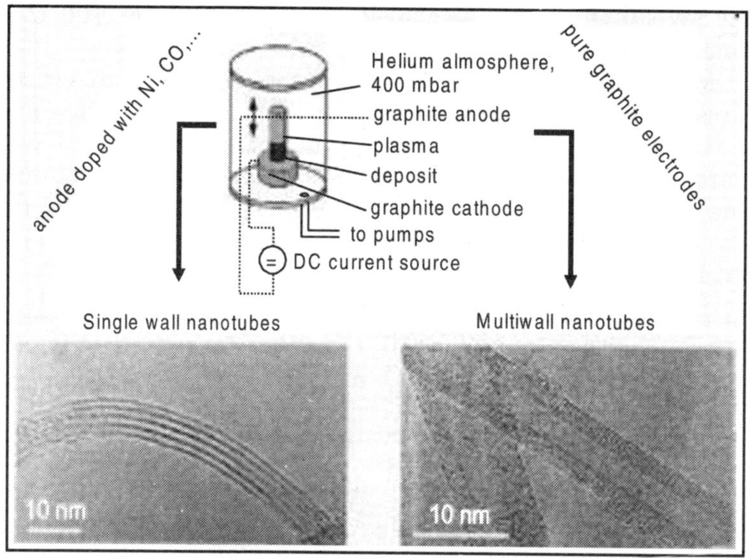

Laser Ablation or Pulsed Laser Vaporization (PLV)

A laser is aimed at a block of graphite, vaporizing the graphite.

Contact with a cooled cooper collector causes the carbon atoms to be deposited in the form of nanotubes.

The nanotube "felt" can then be harvested

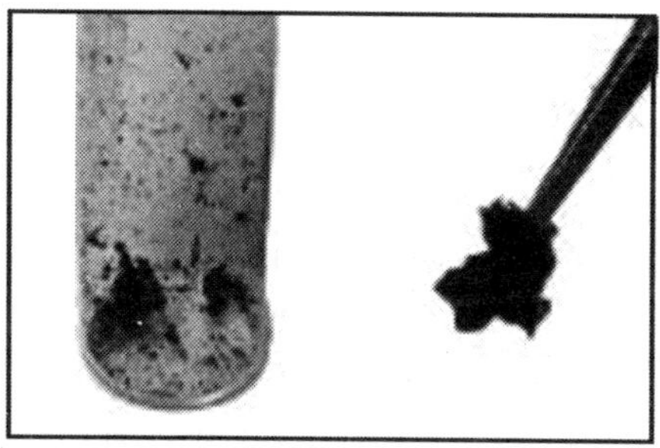

Chemical Vapor Deposition (CVD)

Single-wall nanotubes are produced in a gas-phase process by catalytic disproportionation of CO on iron particles. Iron is in the form of iron pentacarbonyl. Adding 25 per cent hydrogen increases the SWNT yield. The synthesis is performed at 1100 C at atmospheric pressure.

Multi-wall nanotubes are grown in the same apparatus where the catalytic metal particles are supported on a substrate (Si wafers or the quartz furnace tube). Iron is deposited from iron pentacarbonyl or by electron beam sputtering while nanotube growth is achieved by catalytic CVD from hydrocarbon molecules (acetylene, methane) or fullerenes at temperatures between 750 and 1100 C.

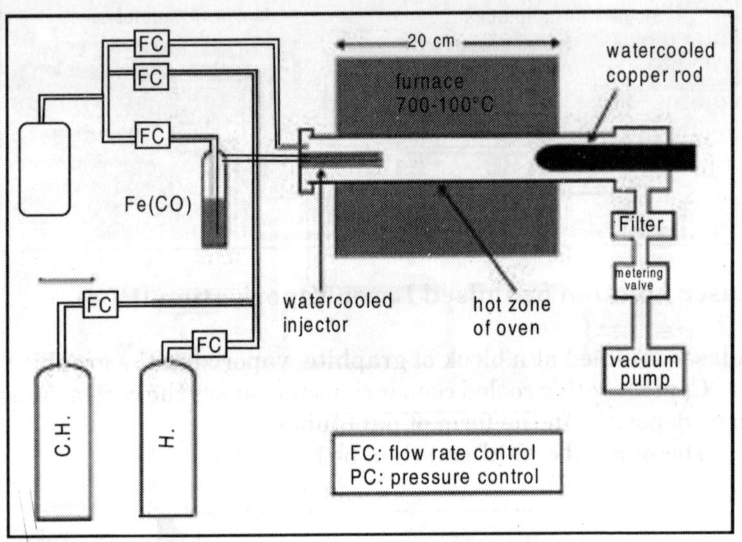

High-Pressure CO conversion (HiPCO)

- Method is similar to CVD
- Carbon source is carbon monoxide
- Catalytic particles are generated in-situ
- Thermal decomposition of iron pentacarbonyl in a reactor heated to 800-1200°C
- High pressure to speed up the growth.
- Bulk production of SWNTs

SCANNING PROBE MICROSCOPIES

The invention of scanning probe microscopies—scanning

tunnelling microscopy (STM), scanning force microscopy (SFM), and a number of more specialised variants—is perhaps the single most important development in the crystallisation of nanoscale science and technology as a new discipline. In these techniques, images are obtained not by gathering reflected or refracted waves from a sample, as happens in conventional microscopies such as light or electron microscopy. Instead, a very fine tip is scanned across the surface of the sample, interacting with it in one of a number of possible ways. The picture is built up electronically by recording the changing interaction with the surface as the tip is scanned across it.

Efforts are already under way to create an interface with a scanning probe microscope that gives the user the sensation of directly manipulating the nanoworld—a so-called virtual reality interface. The aim is to make the sensation of operating the instrument as similar as possible to the way one interacts directly with the physical world.

Scanning Probe Microscopy has been routinely employed as a surface characterization technique for nearly 20 years. Atomic Force Microscopy is the most widely used subset of SPM, which can be used in ambient conditions with minimum sample preparation. AFM is able to measure three-dimensional topography information from the angstrom level to the micron scale with unprecedented resolution.

ATOMIC FORCE MICROSCOPY

Surface Modification

Deliberately creating structures having nanometer sizes is a scientific and technical challenge that has been addressed for several centuries. Clearly, physicists are able to combine atoms, chemists create molecules and biologists create biomolecules with sizes on the nanometer scale. A greater challenge is to engineer structures in the mesoscopic scale of 10-1000 nm. Methods for creating mesoscopic structures include optical lithography, and ebeam lithography. Although successful, these techniques are limited in their use to few researchers because of the expense, greater than 1 million dollars. With the Atomic Force Microscope (AFM), it is now possible to create mesoscopic sized devices for a fraction of the cost of optical and ebeam techniques.

The atomic force microscope is the most widely used scanning probe instrument in nanoscience and nanotechnology. The AFM combines surface profilometry and scanning tunneling microscopy and can operate under ambient conditions on both insulating and conducting materials. In general, the AFM can be used to measure, modify, or manipulate surface structures. This technical note focuses on surface modification—that is, nano-manufacturing processes. The AFM is well suited for R&D and "proof-of-concept" demonstrations of fabrication and patterning techniques in the nanoscale regime.

Surface structures can be modified by a passive probe (e.g. surface indentation) or by several types of active probes. Active probes could employ electrical, chemical, optical, or diffusion processes to change the surface.

Mechanical Surface Modification

The Atomic Force Microscope was initially developed to image the surfaces of insulating materials. However, by accident, it was discovered that the AFM probe could cause mechanical modifications to a surface. Such mechanical modifications can now be used proactively to alter surface topography.

Two types of mechanical surface modifications are possible. In the first, the probe is pushed into the surface. In the second, a line is scratched into the surface. The size of the features depends on the following factors:

- Surface Material (hardness)
- Probe Diameter
- Probe Material (hardness)
- Force of the Probe on the Surface

Electrical Surface Modification

Electrically conducting AFM probes can be used to chemically modify a surface to "draw" an image. For example, applying an electrical bias between the conducting probe and a substrate can locally oxidize selected regions of the surface to form patterns.

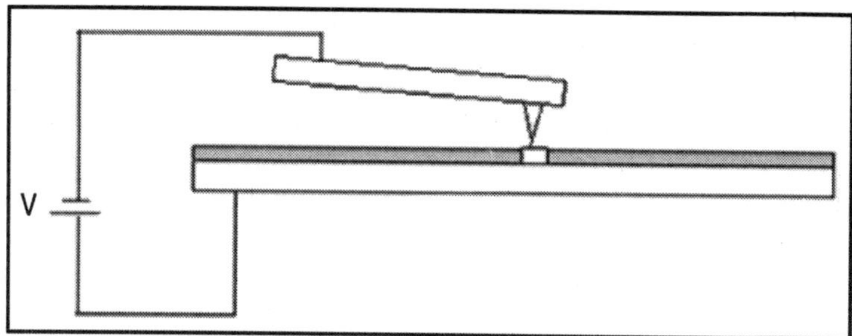

Fig. 4.1: By placing a bias between an electrically conductive probe and a surface, the surface can be modified

The dimensions of the pattern drawn by electrically conducting techniques depends on:

Diameter of the Probe

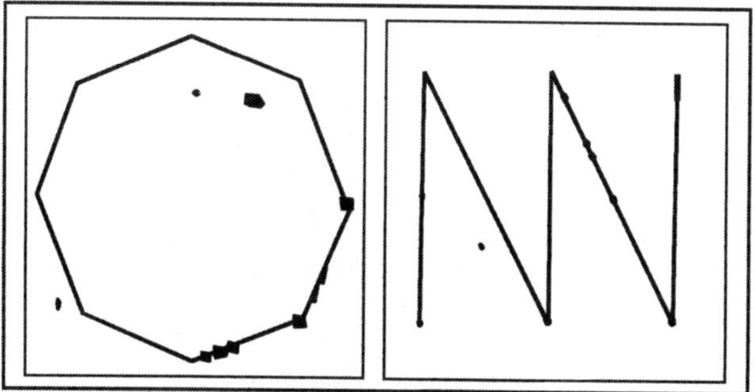

Fig. 4.2: These patterns were drawn with an electrically conductive AFM probe using Anodic Oxidation. At the left the, line widths are 50 nm. At the right, two line widths were drawn using two different biases between the probe and sample

For growing an oxide on silicon, the line-width of the pattern ranges up to tens of nanometers. The thickness can be controlled in the range of 10-50 nm. When the writing is done in ambient air, the line width depends on the relative humidity, because water adsorbed at the tip-substrate interface focuses the electric field and also acts as the anodization medium. The oxide thickness was found to depend on the electric field strength. Fig. 2 shows two patterns written in silicon oxide with an AFM.

Nanolithography by Molecular Deposition

Another way to create a nanopattern is by dip-pen nanolithography™. With DPN™ method, ink molecules are adsorbed on the AFM tip and transferred by diffusion through the liquid meniscus (formed between the probe tip and the substrate surface) onto the substrate. The DPN™ process is patented by NanoInk, Inc. NanoInk is the sole manufacturer of DPN™ process solutions. The technique is illustrated in Figure 3. The liquid can be water or another suitable solvent in which a material (the "pigment") is dissolved. The ink could also be in liquid form and require no solvent. Inks include organic compounds such as alkane thiols, dendrimers, polymers, or large biomolecules such as antibodies, proteins, or DNA. (A dendrimer from Greek dendra for tree is a small, high-molecular weight globular molecule built up from branched units forming a tree-like structure.)

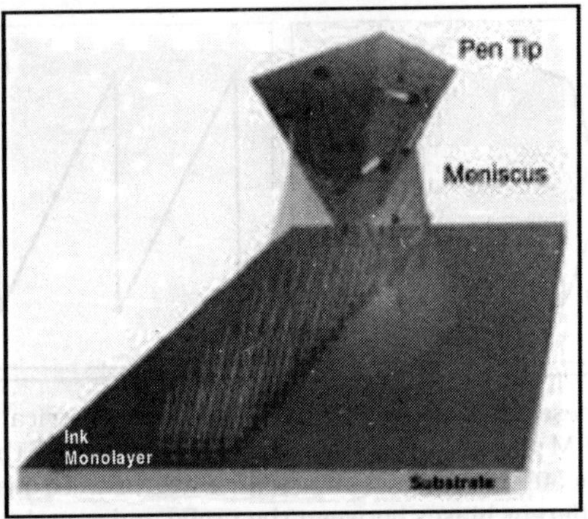

Fig. 4.3: Illustration of the AFM probe being used to deposit molecules on a sample's surface. As the probe moves across the surface, molecules move down the probe and attach to the surface.

Fig. 4.4 shows patterns formed by the DPN™ process on a gold substrate using open loop control of the AFM tip. Note that the box drawn around the LLL pattern is not closed—the open loop control did not allow accurate location of the endpoint.

Advanced DPN™ experiments involve the overlay of patterned

ink layers with the precise deposition of molecules, and this process requires more complex instrumentation. NanoInk has developed a dedicated DPN Writer™ tool called NSCRIPTOR™. Built upon advanced PNI scanner technology, NSCRIPTOR™ offers a sophisticated, user-friendly DPN™ experience with integrated environmental control.

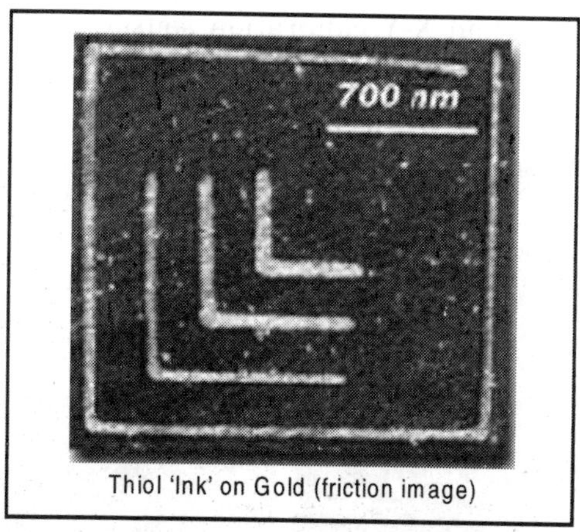

Fig. 4.4: Lateral force microscope image of a pattern written with the DPN™ process. This pattern was written with an AFM that does not have X-Y calibrations sensors. As a result, at the right upper corner of the image two of the lines did not connect as intended.

AFM Lithography Instrumentation

Any atomic force microscope can be used for creating nanometer-sized patterns on a surface. However, the quality and complexity of the patterns depends on specific scanning probe hardware and software. The method of patterning is determined by the types of probes, substrates, and the specific software performance capabilities used to drive the scanning probe tool.

Hardware

The most critical hardware feature that is required is an X-Y calibration system. Because the piezoelectric ceramics used in AFM have unwanted characteristics such as creep and hysteresis, calibration sensors are necessary to guide the motion of the probe. Without the calibration sensors, the probe moves in an unpredictable motion, and it is hard to write complex patterns. As an example, in figure 4 the AFM did not have calibration sensors, and the lines at the upper right of the pattern did not connect as intended. Conversely, Figure 5 is an illustration of a complex pattern drawn with the DPN™ process. Because the instrumentation had X-Y calibration sensors, a higher quality pattern was possible.

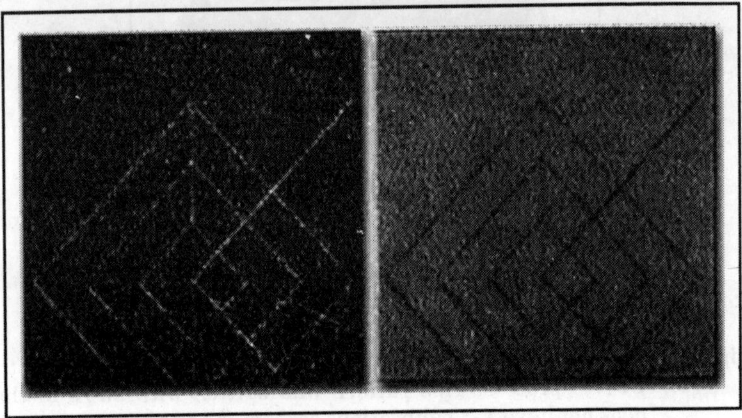

Fig. 4.5: Lateral Force Microscope (LFM) images of patterns created using DPN™. With X-Y calibration sensors the lines at the edges of the boxes intersect as expected. Image written with a NanoInk DPN Writer™ using PNI scanner technology. Line widths vary from 60-100 nm.

Software

Software is used for defining the pattern that will be drawn with an AFM and then for drawing the pattern on a materials surface. The pattern may be drawn on a section of the screen as a combination of dots and lines or it may be imported from a bit-map file. The software allows writing of the patterns by applying a specific force or a specified voltage. Also, the rate of scanning can be specified.

The following is a partial list of the material classes of particles in different environmental media, as analyzed by an AFM:

1.0 *Imaging in Air*

1.1   Dry Powders
1.2   Evaporated Suspensions
1.3   Bio-Particles
1.4   Carbon nanotubes
1.5   TEM Samples
1.6   SEM Samples

2.0 *Imaging in Liquids*

2.1   Bio-Particles in Buffer
2.2   Inorganic Particles 3.2 Hard Surface Materials

3.0 *Imaging Embedded Particles*

3.1   Soft polymer and Bio-materials
3.3   Membranes and defects

Examples of Imaging Nanoparticles with AFM

There are numerous combinations of nanoparticles, substrates and adhesives that have been demonstrated to work with an AFM. Described here are examples of some of the most common types of applications addressed with an AFM.

1.0 *Imaging in Air*

A substantial advantage of using an AFM for Nanoparticle characterization is that the scanning may be done in ambient conditions. Thus, most applications for characterizing Nanoparticles are performed in air.

1.1 *Dry Powders*

A great variety of particles are produced or distributed as dry powders. Commonly used substrates for ultra-fine powder deposition are glass slides, HOPG and mica. In order to increase the adhesive properties of the substrate, poly-L-lysine is may be deposited on the substrate's surface. Once a substrate is chemically treated and dry, it is immediately ready for powder deposition. Powder distribution is achieved by dusting a small amount of powder over the entire area of the substrate, and setting it aside for a few minutes. Then the substrate is flipped over to remove large agglomerate of particles. Dry adsorption works very well for super fine powders, particle size less than 150nm. Deposition rate as well as density of the deposition are the of the challenges associated with this method.

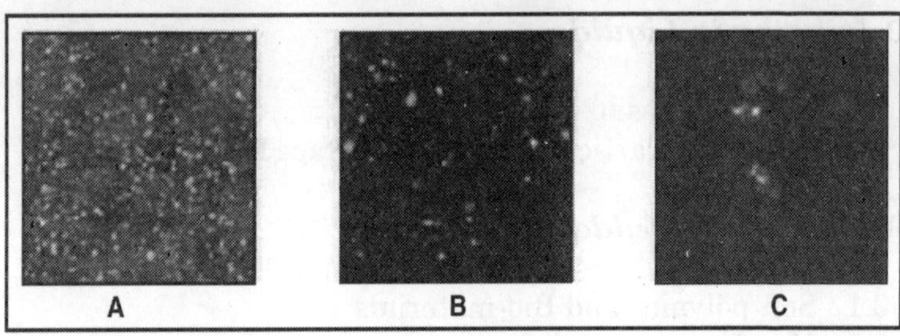

If granular size is larger than 500nm a different method should be used. A polished metal disc works very well as a substrate and thermal wax works well as anchoring medium. A piece of wax is placed on the metal disk and is warmed up on a heating element until the wax softens, at approximately at 60-70 degrees C. After the wax softens, there is a visible liquid interface or its surface. After seeing the liquid interface, remove the metal disc from the heater, wait until the surface just starts to solidify, and sprinkle some powder over the sample area. The sample is ready for AFM imaging when the thermal wax becomes solid and the metal disk is at room temperature, typically after 10-15 minutes. Experimentation is often required to obtain the optimal particle surface density. The depth of the embedding depends on particle weight and size as well as the temperature of the thermal wax, and is very difficult to control with this method. If this effect is undesirable, droplet deposition should be used. In the case of the wet method, the amount of the particle embedded into the anchoring media is negligible.

A novel approach to creating nanoparticles is laser ablation. It is demonstrated that laser ablation may be used for depositing nanosize metal clusters on substrates. In this case there is no additional treatment or preparation required. This method produces a wide distribution of particles size and shapes, depending on the conditions of the irradiating laser.

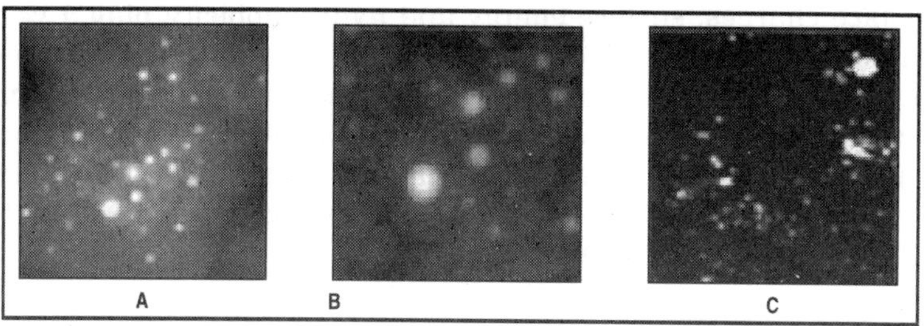

1.2 *Evaporated Suspensions*

Droplet-evaporation or adsorption methods are used for preparing AFM samples from liquid suspensions. A droplet of liquid is deposited on freshly cleaved mica or a poly-l-lysine covered slide. The droplet is then carefully washed after allowing the sample to sit for about 10 minutes. To dry the sample before scanning, either leave it overnight in a dust protected environment or use a furnace/heater to accelerate the drying process.

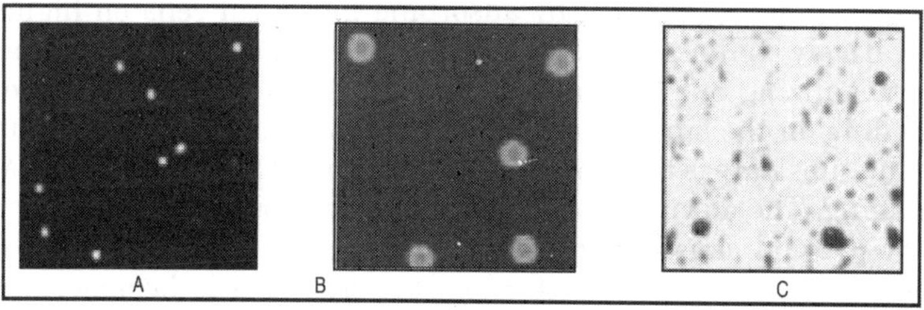

Certain kinds of particles, quantum dots for example, come in a toluene solution. If the solution or suspension comes in any nonaqua form, it is very important to choose the substrate accordingly. Glass or silicon work well for toluene, Figure C.

1.3 *Bio-particles*

Using the correct sample preparation techniques for life-science and biological applications is extremely critical because an immobilized specimen can degrade during sample preparation or even during imaging. Requirements for substrate flatness, chemical compatibility and reagent purity are rigorous. Also, surface charges, surface energy and hydrophobicity play a very important role in selecting the optimal sample preparation methodology. There are several review papers written on biological sample preparation. There are the major methods: absorption, replication, and mechanical trapping. Sometimes additional fixation is necessary and several methods can be combined to achieve the desired results,

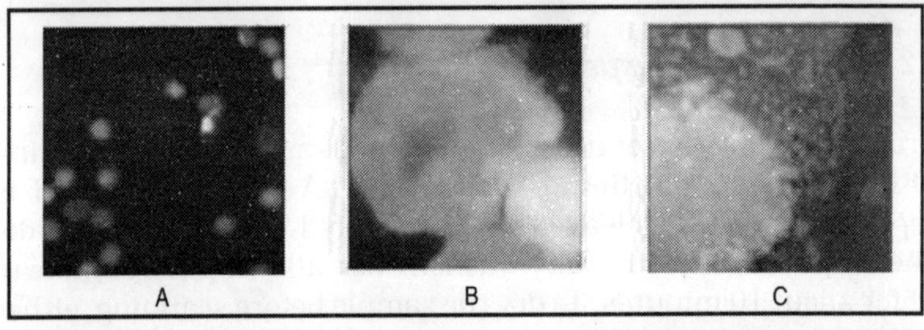

Physorption (physical absorption) or non-covalent methods, for example aerosol-spray deposition, immersion and droplet-evaporation, is achieved by adsorbing biological cells on highly negatively charged mica. Additional chemical treatment, such as fuctionalizing by salinization, can be used to facilitate stronger bonding on the surface of the biological specimen. Tight affinity to the substrate is the mandatory requirement for successful AFM imaging. The downside of the non-covalent absorption method is that it could cause undesired re-arrangements of the bacterial cells. If displacement or distortions are critical or if the molecular object has to be integrated into a complex molecular assembly, then covalent methodologies should be used.

Sometimes fixing with glutaraldehyde is necessary to minimize tip-object interaction and to prevent possible damage of the biological sample. Studies show that mechanical trapping of biological objects in a membrane filter appears to be the most reliable method to measure actual surface topography. It is beyond the scope of this book to describe all existing techniques.

In the case of many pharmaceutical applications, particles come either as dry powders or liquid suspensions. If this is the case, sample preparation for AFM is the same as for inorganic powders and suspensions as described in Powder/Suspensions, sections.

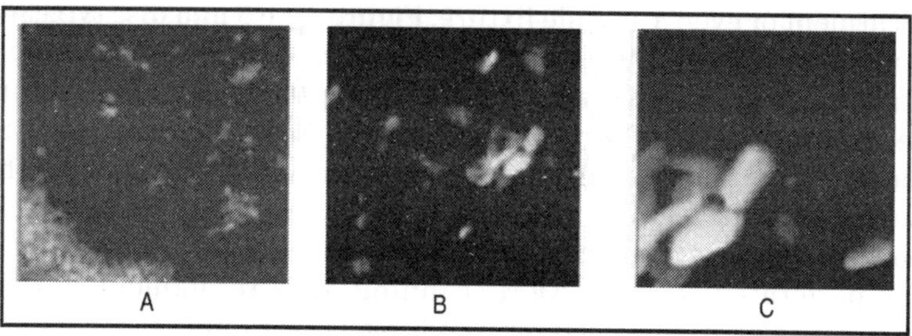

1.4 Carbon-nanotubes

Carbon nanotubes, nanowires and whiskers are a subset of nanoparticles. These particles are normally produced in large quantities as powders or are grown directly on a substrate. Arc-discharge, laser ablation and chemical vapor deposition (CVD) methods have been successful in making carbon fibers, filaments, and nanotube materials. These methods are well described by H. Dai. Typically one of two methods is used for preparing nanotube samples for AFM imaging: catalyst growth or deposition. Catalyst growth is the best method for creating a clean sample for studying the unique properties of single-wall nanotubes.

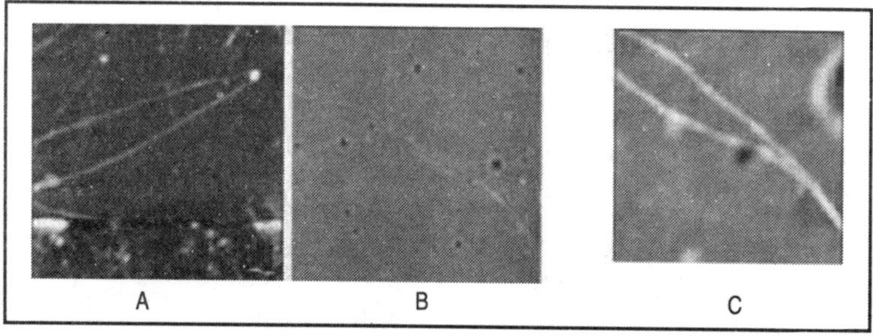

When preparing carbon nanotube samples for AFM imaging with deposition, it is important to use a dispersant. Very diluted dispersant suspensions of carbon nanotubes are spin coated on a silicon wafer, rinsed thoroughly with water, then dried in air. Any commercial spin-coater may be used.

1.5 *TEM-samples*

Imaging of samples prepared for TEM analysis is very simple and straightforward with AFM. There is no need for additional surface treatment or even a sample fixture, Figure. The 3 mm disc typically used for TEM analysis must be firmly fixed in the AFM sample stage. Usually double sticky tape or carbon tape works well to secure the disc. The perforation in the TEM sample can easily be located with the optical microscope in the AFM stage. Once the perforation is located, the AFM probe can then be precisely positioned over the electron transparent area or father away from the perforated area for AFM scanning. It is recommended that vacuum tweezers be used for handling TEM samples.

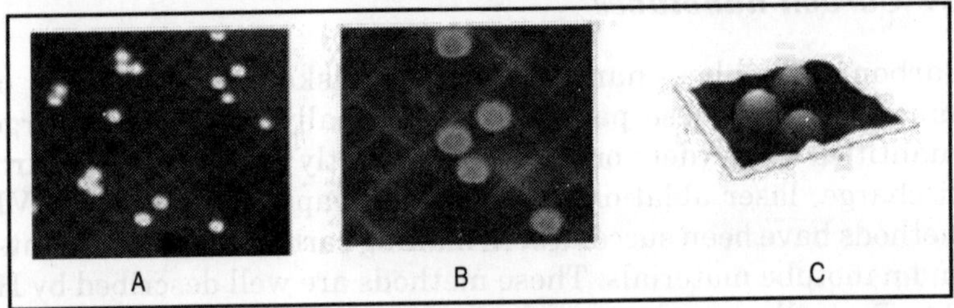

1.6 *SEM-samples*

Samples made for SEM imaging are may be directly imaged in an AFM, Figure 11. Sample preparation procedures can be simplified for AFM scanning because the sample does not require a conductive coating. When compared to an SEM, traditional considerations for good material contrast are not important. For example, in an AFM precipitated particles on a flat substrate will always appear with good contrast regardless of material choice for the particle and sub-strate.

2.0 *Imaging in Liquid*

AFM is an essential tool to identify topographical features of particles submerged in liquid. The range of applications include soft polymers, bio-particles (cells, membranes, viruses) and a variety of inorganic particles.

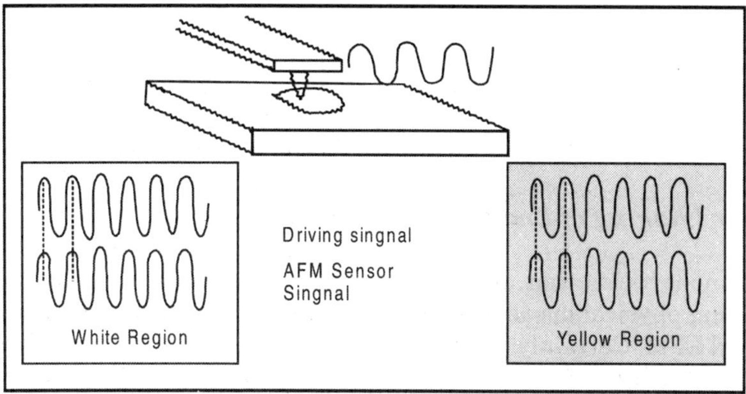

2.1 *Bio-particles in Buffer*

An important advantage of AFM over other microscopy techniques is the ability of an AFM to image biological samples in a native aqueous environment. AFM offers the possibility of in-vivo monitoring of the dynamics of biological changes in living cells, viruses and micromolecular crystals. Imaging in liquid with an AFM requires a stable immobilization of biological objects. Absorption on a polycationic treated surface or on an agarose coating provide stable fixation for experiments in liquid. Absorbing specimens directly from the buffer solution can be controlled by the electrolyte concentration and pH of the buffer solution. Glutaraldehyde fixing is necessary for certain applications in a bio-AFM sample. The fixing agent is applied after absorption. In fact, fixation destroys molecular functionality and could affect true structure. Hydrophobic substrates and bio-incompatible agents are not recommended for solutionbased measurements and should be avoided.

Both contact or close contact imaging modes can be used in liquid. Close contact mode is the mode of choice for imaging soft samples in air, however in liquids it may not be the best technique. In fact, the contact mode is the optimum mode for imaging samples with large contact area, such as purple membrane. In the case of samples with individual particles attached to the substrate, other imaging modes are used

2.2 *Inorganic Particles in Intermittent Medium*

Both the particle and adhesive holding the particles in place during imaging should be un-dissolvable when imaging in liquid for obvious reasons. It is also very desirable if both adhesive and particles are hydrophilic if imaging is done in an aqueous solution.

3.0 *Imaging of Embedded Particles*

Nanoparticles that are imbedded in surfaces can be visualized using physical measurement techniques such as vibrating phase and LFM. Certainly, if particles largely extend out of the surface, then traditional methods for topographic imaging will work.

3.1 *Soft and Bio-materials*

Often it is desirable to image particles embedded in a solid medium, bio-tissue, or polymer thin film. In the case of relatively soft material like organic tissue or soft polymers it is important to cross section the specimen and make a very smooth, clean cut. A microtone is typically used to produce 0.5 micron or thinner slices that are suitable for AFM imaging, Fig.-A. The microtone slices must be firmly fixed on a glass substrate before AFM scanning. Chemical etching of semi-thin sections of an epoxy-resin embedded specimen is a very good technique for visualizing embedded particles with an AFM. Being able to scan with an AFM in this case depends on sample preparation before the particles are embedded in the substrate.

Thin films can be spin coated on silicon or a glass substrate and the examined with an AFM with no additional treatment, Fig. 13-B. If a composite material is the subject of investigation and the constituent component have dramatically different elastic, adhesive or frictional properties, then material contrast using vibrating phase modes can be achieved. Vibrating phase imaging can provide unique information about local materials distribution for composite/organic thin film, Figure C. Phase-lag data are obtained simultaneously with topography data in AFM, Figure

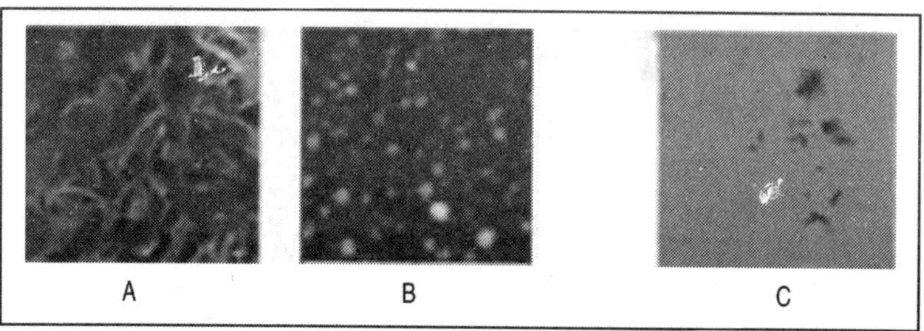

3.2 Hard Materials

If a matrix material is rather hard, for example metal or ceramic, the standard polishing/etching technique employed in SEM may be used for AFM scanning, Figure. There are numerous combinations of nanoparticles, substrates and adhesives that have been demonstrated to work with an AFM. Described here are examples of some of the most common types of applications addressed with an AFM.

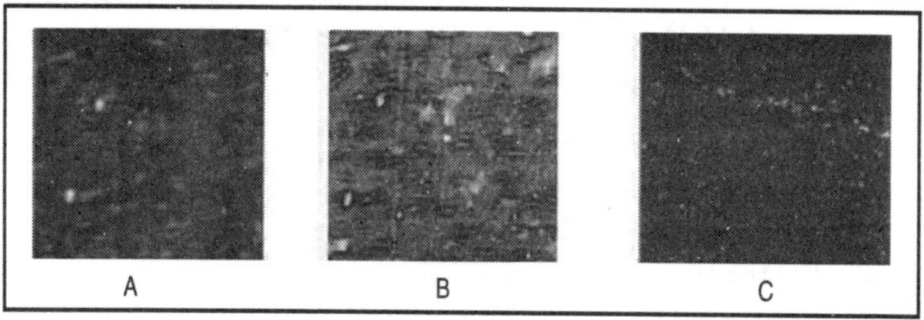

3.3 Membranes and Defects

Porous materials and materials with cavities/voids are considered the same as particle specimens, Figure. In the case of membrane defect visualization, there is no extra sample preparation required. For bulk materials, AFM sample preparation can be done as described in earlier section.

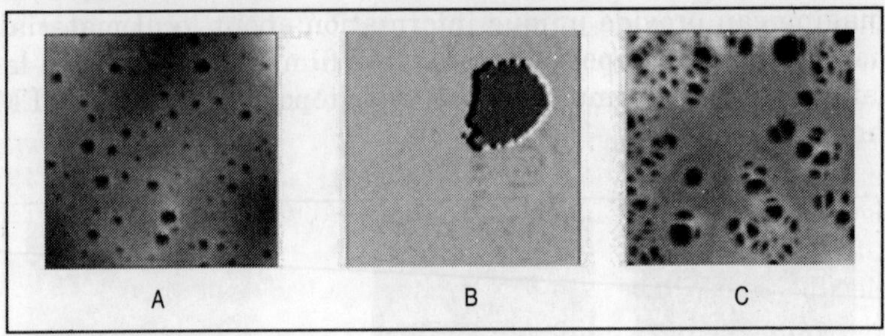

Conclusions

Sample preparation techniques for powders, colloids/suspentions, bio-objects and embedded particles are outlined in this chapter. Examples of procedures along with AFM images illustrate material science, pharmaceutical or biological applications. AFM sample preparation is derived from traditional optical/SEM/TEM methods,and is much simpler and less time consuming. Nevertheless, the importance of adequate surface treatment can not be overemphasized.

Speeding up the process of obtaining vital data is critical for many applications and definably makes AFM more attractive for individual particle imaging. In general, AFM individual particle size characterization is both cost and time effective. AFM resolution is greater or comparable to traditional techniques. The main advantage of AFM for particle characterization is unambiguous morphology determination along with direct measurements of volume and 3D display.

Advantages of Using an AFM for Particle Analysis are:

- Faster than SEM and TEM
- Instrumentation is more affordable
- Direct three dimensional map
- Works on many types of nanoparticles

Single Molecule Techniques

Chemistry and biochemistry deal with the properties of molecules, but almost invariably experiments are done on very large collections of them. A gram of material contains about ten trillion billion atoms. New techniques—including scanning force microscopy—are capable of interrogating the properties of single molecules.

The optical tweezers technique, in which a molecule is attached to a micron size bead held in the focus of a powerful laser, allows single molecules to be moved around, stretched and deformed. In addition to providing information about the properties of single molecules that will be essential in the design of nanoscale devices, these experiments challenge a prevalent assumption in many branches of science that the properties of an ensemble of molecules are dominated by the molecules with average behaviour.This may true when one has a test-tube or tank reactor full of molecules, but in a nanoscale system, such as a single cell in biology, an important molecule may be present in rather small numbers, and individual molecules which behave in a way that departs from the average may play a disproportionately important role.

Microlithography and Mems

The tools with which sub-micron structures are made for electronic devices are now mature and highly optimised. The process involves laying down thin layers of material, putting a pattern on it and selectively removing material to develop the patterns.These make up the technologies of microfabricationThe technologies developed for the electronics industry have also been adapted to make miniaturised mechanical and optical systems from silicon. Some of these products are already commercialised, for example sensors in air bags and arrays of micron-sized mirrors, which can be individually moved around to steer arrays of light beams in optical communication applications.The extent to which these devices can be shrunk into the nanoscale, however, depends on fundamental physical limits, such as the prevalence of sticking and friction.This technology is also the basis for the production of tips for scanning probe microscopes, underlying the global semiconductor industry.

Electron Beam Lithography and Focused Ion Bom Bardment

Laboratory-based tools exist which permit the extension of top-down manufacturing techniques to nanoscale dimensions.

Conventional optical lithography—in which a resist polymer is exposed by being illuminated by light—is limited by the wavelength of the light, which determines the sharpness of the features that can be duplicated from the mask. The smallest feature size that can be drawn with optical lithography has been reduced because lasers have become available with shorter and shorter wavelengths, moving into the ultra-violet region of the electromagnetic spectrum. There seem to be, however, physical limits on the smallest wavelengths that it is possible to generate with conventional laser designs.

An alternative approach to lithography uses a beam of electrons rather than a beam of light to develop the pattern. Just as an electron microscope has a higher resolution than a light microscope because the wavelength of electrons is substantially smaller than that of light, so a finely focused beam of electrons can be used to pattern a resist on a much finer scale than light. This technology, known as electron beam lithography, has been available for around 30 years but its impact has been limited by its expense, and perhaps more fundamentally, by the fact that it is a serial rather than a parallel process. Rather than creating an entire pattern in one shot, as an optical lithography process does, each line in the pattern has to be drawn individually. This greatly reduces the speed of the process and the number of devices that can be made by it. Another approach to nanofabrication uses a beam of charged atoms—or ions—which can physically shape a sample on the nanoscale, just as a milling machine can shape a macroscopic piece of metal. Again, this is a serial process capable of making only one object at a time.

Thin, Precise Coating Technologies—MBE and CVD

Much of the emphasis of current semiconductor nanotechnology has been on controlling the structure of semiconductors on the nanoscale—typically by making devices consisting of alternating, very thin, layers of different semiconducting materials—to create new composite materials with designed electronic properties.

This has already led to substantial economic impacts through the development of new lasers, light emitting diodes, and other optoelectronic devices. Very similar principles can be used to make new magnetic materials which allow the very much more sensitive transduction of magnetically stored information.

Soft Lithography

The techniques that have evolved for manipulating matter on the nanoscale from the computer and optoelectronics industry have very high capital and running costs and need specialised expertise to make use of them.This has greatly limited the rate at which they have been adopted outside the semiconductor world, either in academia or in industry. One interesting recent development has been the introduction of new techniques to pattern surfaces which are ultimately less effective than conventional lithography, but which are orders of magnitude cheaper.These techniques, collectively known as soft lithographies, rely on advances in surface chemistry, which allow one to create well-ordered layers a single molecule thick on easily available substrates, like evaporated layers of gold (alkyl thiol self-assembled monolayers).

Simple printing techniques using soft elastomers allow surfaces to be patterned with these molecules on a sub-micron scale using cheap and easily available equipment.These developments have allowed branches of science and technology, such as tissue engineering, the branch of biomedical engineering concerned with the creation of new skin and organs, that would not normally be involved in conventional nanofabrication, to move into this area, and also offer the potential for cheap manufacturing routes to any products that are developed.

Software—Molecular Modeling

Molecular modeling is techniques used to build, display, manipulate, simulate and analyze molecular structures, and to calculate properties:

1. *Molecular mechanics* methods take a classical approach to calculating the energy of a structure.
2. *Molecular dynamics* can be used to simulate the thermal motion of a structure as a function of time, using the forces acting on the atoms to drive the motion.
3. *Quantum mechanics* takes account of conjugation (quantum electron orbital effects).

Let Us Study Various Tools and System Available for Software Modelling

Amorphous Cell

Amorphous Cell is a suite of computational tools that allow scientist to construct representative models of complex amorphous systems and to predict key properties. By observing the relation between system structure and properties, scientist can obtain a more thorough understanding of the important molecular features, allowing you to better design new compounds or new formulations. Among the properties that scientist can predict and investigate are cohesive energy density, equation-of-state behavior, chain packing, and localized chain motions.

The behavior of amorphous materials is critical to products including plastics, glasses, foods, and chemicals. Researchers studying amorphous polymers, for example, seek to optimize their mechanical behavior, the transport of molecules through the system, and their surface and interface interactions. These properties impact the polymer's performance in applications including separation processes, packaging, and in drug delivery

CASTEP

CASTEP is quantum mechanical program employing density functional theory (DFT) to simulate the properties of solids, interfaces, and surfaces for a wide range of materials classes including ceramics, semiconductors, and metals. First principle calculations allow researchers to investigate the nature and origin of the electronic, optical, and structural properties of a system without the need for any experimental input other than the atomic number of mass of the constituent atoms. CASTEP is thus well suited to research problems in solid state physics, materials science, chemistry, and chemical engineering where researchers can employ computer simulations to perform virtual experiments which can lead to tremendous savings in costly experiments and shorter developmental-cycles.

COMPASS

COMPASS stands for Condensed-phase Optimized Molecular

Potentials for Atomistic Simulation Studies. It is the first forcefield that has been parameterized and validated using condensed-phase properties. Consequently, this force field enables accurate and simultaneous prediction of structural, conformational, vibrational, and thermophysical properties for a broad range of molecules in isolation and in condensed phases, and under a wide range of conditions of temperature and pressure.

Currently, the coverage includes the most common organics, inorganic small molecules, polymers, some metal ions, metal oxides and metals. All of the parameters in COMPASS are derived in a consistent manner so that, in principle, one can study very different systems including interfaces and mixtures.

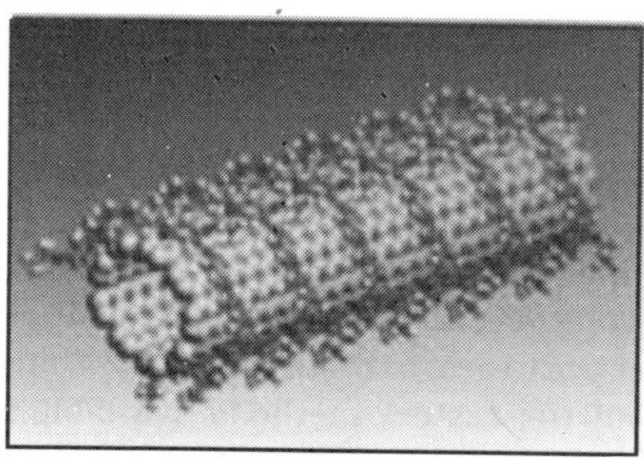

A polymer chain helically wrapping a carbon nanotube. Structure optimized with MS Modeling's Discover and the COMPASS forcefield.

Discover

Discover® is MS Modeling's 'simulation engine'. It incorporates a broad spectrum of molecular mechanics and dynamics methodologies that have demonstrated applicability to molecular design. Using a carefully-derived forcefield as the foundation, minimum energy conformations, as well as families of structures and dynamics trajectories of molecular systems, can be computed with confidence.

Discover provides underlying calculations for products such as Amorphous Cell. It also supports varied simulation strategies. Periodic boundary conditions allow the simulation of solid-state systems, whether crystalline or amorphous, and of solvated systems.

Comprehensive analysis features enable the extraction of pertinent results from the simulation.

A single strand of DNA attached to a Gold surface. The system is a simplified prototype of a DNA chip. Structural and dynamical properties can be studied using MS Modeling's Discover

DMol3

DMol3 is a premier, robust DFT program with a long track record of successful commercial applications. Owing to its unique approach to electrostatics, and an efficient optimizer, DMol3 has long been one of the fastest methods available for performing molecular DFT calculations, especially for large molecules. It now extends this approach to solid-state problems via a high-performance internal coordinate optimizer that can treat the full range of solid-state state systems. The DMol3 code delivers the first robust, general-purpose internal coordinate optimization scheme for periodic systems. A new transition state search employing a combination of LST/QST algorithms with subsequent conjugate gradient methods greatly facilitates the optimization of the transition state structure. This robust and easy-to-use scheme works for both molecular and periodic systems and offers a significant speed-up as compared to traditional methods. The transition state toolbox contains the powerful transition state confirmation method Nudged Elastic Bands (NEB).

NEB enables scientists to verify that the transition states obtained by the LST/QST technique is the one linking the reactants and the products.

The Handy-Tozer functionality enables the simulation of hydrogen bonded systems and accurate molecular structures and thermodynamic data can be obtained. Using parallel versions of the DMol3 code large problems can be tackled.

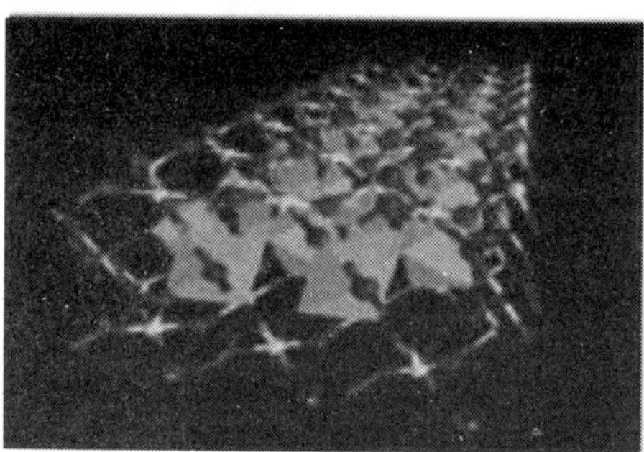

SnO$_2$ **nanoribbon with exposed (10-1) and (010) surfaces. Such a system can be used as ultrasensitive nanosensors for various gases, e.g., NO$_2$, O$_2$ and CO. Calculations performed with MS Modeling's DMol3.**

DPD

DPD is a robust, coarse-grained, dynamics algorithm for simulating soft matter. The basic idea of DPD is that it should be possible to replace 'blobs' or 'droplets' of fluid with individual particles, which interact in such a way as to reproduce Newtonian hydrodynamics of the fluid as a whole. Thus, the contribution of the microscopic details of the system, i.e., the main concern of atomistic simulation methods, is integrated out, allowing access to much larger time and length scales. The underlying chemistry of the system is not lost in the coarse-graining. The resulting 'beads' interact with one another via a pair-potential related to the Flory-Huggins interaction parameter. This parameter can be measured experimentally or obtained from modeling with atomistic detail.

The particles are subject to pair-wise forces, which comprise: soft repulsions, stochastic noise and dissipation (conceptually similar to a viscous drag). Navier-Stokes behavior is obtained with a suitable choice of functional form of these potentials and so the method captures hydrodynamics.

A wide range of properties can be calculated with DPD, including:

Phase morphology End to end distance distribution of any chain molecules.

- Stress tensor
- Surface tension
- Critical micellar concentration
- Aggregation and coagulation
- Effect of confinement on miscibility
- Effect of shear on morphology
- Concentration profiles
- Diffusion rates of various species.

Forcite

Forcite calculates single point energies and performs geometry optimization, i.e. energy minimization, of molecules and periodic systems. For periodic systems, Forcite allows the optimization of the cell parameters simultaneously with the molecular coordinates. In addition, an external hydrostatic pressure may be applied.

During a geometry optimization of a crystal structure, Forcite preserves the symmetry defined by the space group.

Since Forcite is a molecular mechanics tool, its calculations draw upon forcefields, and associated parameters settings. Forcite is designed to work with a wide range of forcefields, and give easy and flexible access to the associated parameter options.

The image below shows hydrogen interaction with a tungsten surface. Geometry optimization with Forcite calculates the physisorption of the hydrogen molecule, which can be used as a starting structure for chemisorption simulations with DMol3 or CASTEP.

Reflex Plus

The Reflex Powder Indexing tool determines the cell parameters and crystal system by indexing the experimental powder diffraction pattern.

The modified Pawley procedure available in the Reflex Powder Refinement module refines the cell parameters, peak shape, and background parameters, and is also a helpful tool to narrow down the list of possible space groups.

The Powder Solve method performs a search of possible arrangements and conformations of the molecular fragments in the unit cell. It finds a structure whose simulated powder pattern matches the experimental one as closely as possible and is also chemically viable. A final refinement of the proposed solution is performed with the rigid-body. Rietveld Refinement functionality available in Reflex.

VAMP

VAMP allows fast and reliable calculations to be performed on large systems utilizing semiempirical methods. The tasks that can be performed by VAMP range from geometry optimization, transition state search and optimization to the evaluation of many chemical and physical properties. VAMP can optimize monomers and oligomers, calculate input charges for molecular dynamics simulations, for UV/VIS spectra calculations, to calculate potential energy surfaces for force field parameterization, and for IR spectra calculations where force-fields are not applicable.

VAMP implements the Natural Atomic Orbital-Point Charge (NAO-PC) model for molecular electrostatic properties. It gives accurate dipole, quadrupole, and higher moments, and high quality Molecular Electrostatic Potentials (MEPs), many times faster than comparable methods. Molecular quadrupoles calculated with MNDO, AM1, or PM3 agree at least as well with experiment as those calculated using ab initio theory at the MP2/6-31G* level.

VAMP can successfully optimize geometries for which other semi-empirical programs cannot find the minimum. The program also contains two different transition state optimizers: eigenvector following and Powell's method.

Solvent effects are simulated using numerical Self-Consistent Reaction Field (SCRF) calculations for ground and excited states, and Conductor-like Screening Model (COSMO) for ground states.

VAMP is the only program that calculates both ESR hydrogen hyperfine coupling constants and 13C chemical shifts using artificial neural nets. Current accuracy (standard deviations from experiment) is about 0.5 Gauss for ESR coupling constants and 6-8 ppm for 13C chemical shifts.

VAMP provides a number of molecular properties such as ionisation potential, multipole moments, accurate molecular polarizabilities, atomic polarizabilites, and optical spectra.

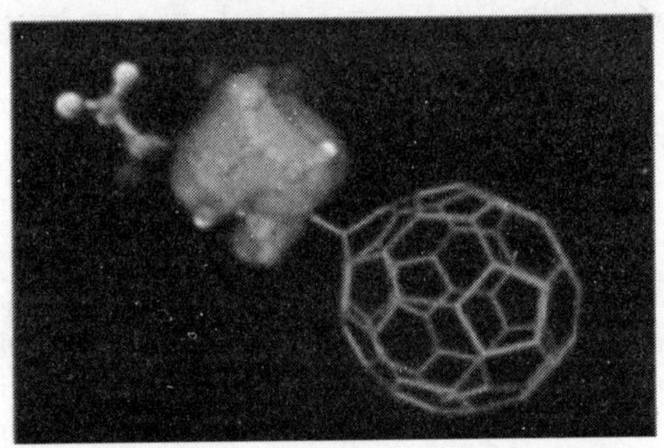

'Buckyaspirin'-an aspirin molecule attached to a Buckyball. An orbital localized on the phenyl ring is displayed. Calculation performed with MS Modeling's VAMP.

Materials Visualizer

Materials Visualizer provides modeling, analysis and visualization tools. Combined with its clear and intuitive graphical user interface, these offer a high-quality environment into which scientist can plug any MS Modeling product.

Run stand-alone, Materials Visualizer offers a comprehensive modeling and visualization system. You can increase your understanding of materials and improve your communication of chemical information. Visualize and model structures from chemical databases. Build your own molecules and materials with fast, interactive sketching, building, and editing tools. Calculate and display key structural parameters. Display the results of calculations from MS Modeling computations—animated dynamics trajectories, graph data, and molecular models. Annotate models for pictures and diagrams and produce exceptionally high quality hard-copy output.

MesoDyn

MesoDyn takes a coarse-grained description of a complex fluid and performs time-evolution dynamics of the density and potential fields of the system. The coarse-graining of the system involves replacing polymer chains by a Gaussian representation with the same response functions and including non-ideality of the system via effective external potentials, the magnitude of which is determined by the Flory-Huggins interaction parameters of the various binary pairs in the system. Electrostatics may be included via the Flory-Huggins parameter, or may be explicitly included for each bead in the system.

The dynamics of the system is described by a set of so-called functional Langevin equations. In simple terms these are diffusion equations in the component densities, which take account of the noise in the system. By means of numerical inversions, the evolution of the component densities is simulated, starting from an initially homogeneous mixture in a cube of typical size 100-1000 nm and with periodic boundary conditions.

A wide range of properties can be calculated with MesoDyn, including:

- Phase morphology
- Aggregation and coagulation
- Effect of confinement on miscibility
- Effect of shear on morphology
- Concentration profiles
- Density histograms
- Free energy and entropy evolution
- Species potential plots (indicating areas of high and low energy)
- Compositional order parameters, giving a direct measure of phase separation.

Nanomaterials, especially nanotubes of various kinds as well as nanodots, exhibit unique combinations of properties that make them prime candidates for a range of device applications. However, conventional device models tend to fail, since at the nanoscale, electrons no longer flow through electrical conductors like rivers—conventional physics and 'water-through-a-pipe' modeling thus no longer applies. At this scale quantum mechanical modeling tools are required. For example, MS Modeling's DMol3 ideally combines the efficiency and accuracy needed for such investigations. In particular, its new multiple point capability allows for the efficient study of infinite nanotubes and the transition state searcher facilitates the study of surface chemistry which is known to modify the conductance and field emission properties. Accuracy and efficiency can be conveniently tuned by using a real space cut-off radius allowing for simulations of very large structures. Let us see how these tools are applied in some topical areas of device and nano-electronics development.

Opto-electronics

Oxygen Manipulation of the Structural and Optoelectronic Properties of Silicon Nanodots

Researchers have used MS Modeling's CASTEP to study the role of oxygen on the structural and optoelectronic properties of silicon nanodots.

Such an understanding will enable these properties to be manipulated, leading to commercially viable nanoscale solid-state lighting devices—a major commercial application of nanotechnology.

Nanodots, also known as quantum dots, consist of 100s-1000s of atoms of inorganic semiconductor nanoparticles and are approximately one billionth of a meter in size. Developed in the mid-1980s for optoelectronic applications, they have interesting structural, electronic, and optical properties—they strongly absorb light in the near UV range and re-emit visible light that has its color determined by both the nanodot size and surface chemistry. And as the size of nanodots can be controlled during synthesis with nanoscale precision, so the optical properties can be manipulated. In addition, nanodots have a longer life than organic fluorophores, and have a broad excitation spectrum. These factors combined make the use of quantum dots as light-emitting phosphors a strong candidate for a major application of nanotechnology in the future Silicon nanodots have, in particular, have emerged over the last 10 years as a hot area of research due to the fact that a reduction in size of this semiconducting material to the nanometer scale dramatically alters their physical properties. In addition, the 1990 discovery that porous silicon exhibits photoluminescence properties, has led to a flurry of research activity, with commercially viable solid-state lighting devices made from Si nanostructures seemingly within reach.

As porous silicon reacts with the atmosphere, leading to major structural (and thus optical properties change), the theoretical role of different passivation species (to passivate is to coat (a semiconductor, for example) with an oxide layer to protect against contamination and increase electrical stability) must be fully understood.

With the knowledge that silicon nanocrystals dispersed in SiO_2 show an optical gain researchers at the University of Modena and Regio Emilia, Italy, used MS Modeling's CASTEP to study the role of passivating oxygen on the structural and optoelectronic properties of silicon nanodots .

Marcello Luppi and Stefano Ossicini used denisty functional theory (DFT) to investigate:

The changes in the optoelectronic properties when O is absorbed onto hydrogenated Si nanocrystal. The different role played by single and double SiO bonds

How a SiO_2 matrix influences the physical properties of a Si nanocrystal.

Using CASTEP, the Scientists Showed

In hydrogen covered Si nanocrystals single-bonded oxygen atoms lead to small variations in the electronic properties yet large changes in structure. However, double-bonded oxygen atoms lead to small geometry variations yet a large energy gap reduction, explaining the huge photoluminescence red shift observed in high porosity silicon after oxygen exposure.

The studies on Si nanocrystals embedded in a SiO_2 matrix revealed:

The prescence of the nanocrystals only slightly deforms the SiO_2

cage determining the formation of an interface region of stressed SiO_2 between the nanocrystals and the matrix

New electronic states originate in the SiO_2 band gap

Both nanocrystal Si atoms and interface O atoms affect optical properties.

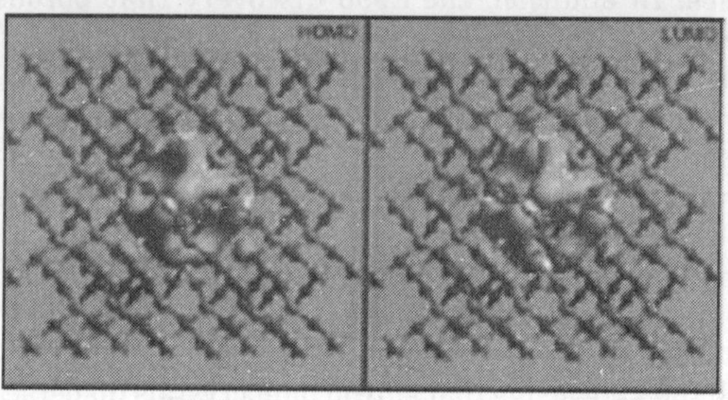

The HOMO and LUMO isosurfaces at fixed value show that the distribution is totally confined in the Si NC region with some weight on the interface O atoms. These dot-related states originate strong absorption features in the optical region. These features are entirely new and can be at the origin of the photoluminescence observed in the red optical region for Si nanocrystals immersed in a SiO_2 cage.

These findings help to explain experimental optical property observations, and should lead to the fine tuning of nanodot optical properties—paving the way to commercially viable solid-state lighting devices based on nanodot technology.

The use of CASTEP enabled scientist to perform a first principle study on semiconductor nanodots embedded in an insulator host matrix in which for the first time the whole system has been geometrically optimized with the performance of the code scientists are able to handle hundreds of atoms and to study the electronic and optical properties of the system in a very accurate and efficient way. The graphic user interface gives the perfect tool for drawing models and for analyzing the results.

Doping

Researchers have used MS Modeling's CASTEP to study the effect of nitrogen substitutional impurities on the electronic properties of single-wall carbon nanotubes.

Such an understanding will enable the electronic properties of carbon nanotubes to be fine tuned. This should lead to the design of better electronic devices, leading to the use of carbon nanotubes in many nanotechnologies and molecular electronics. Carbon nanotubes are long, thin cylinders of bound carbon atoms, about 10,000 times thinner than a human hair, and can be single- or multi-walled. They have remarkable electronic and mechanical properties that depend on atomic structure and more precisely on the manner in which the graphene sheet is wrapped to form a nanotube (chirality). They can vary from being metallic to semiconducting.

Carbon nanotubes are a hot research area, fuelled by experimental breakthroughs that have led to realistic possibilities of using them in a host of commercial applications: field emission-based flat panel displays, novel semiconducting devices in microelectronics, hydrogen storage devices, chemical sensors, and most recently in ultra-sensitive electromechanical sensors. As a result they represent a real-life application of nanotechnology.

However, two major challenges remain an obstacle to the full commercialization of nanotube-based nanotechnologies and molecular electronic devices:

The manipulation of individual tubes is difficult owing to their size, and The ability to manipulate nanotube properties to suit the application has to be achieved.

In semiconducting nanotubes, introducing impurities, a process known as doping, is the main method of tuning properties to make electronic devices. Doping is also a way of creating chemically active impurity sites.

Using CASTEP, the researchers found that, at low concentrations of nitrogen impurity (less than 1 atom%), the impurity site becomes chemically and electronically active. In addition, the team found that an inter-tube covalent bond can form between neighboring nanotubes with impurity sites facing each other. •

The effect of nitrogen doping in two zigzag nanotubes. The left image shows the charge density, the right image shows the density of the HOMO orbital. The chemical bond is formed between the two carbon atoms that have the maximum spin density.

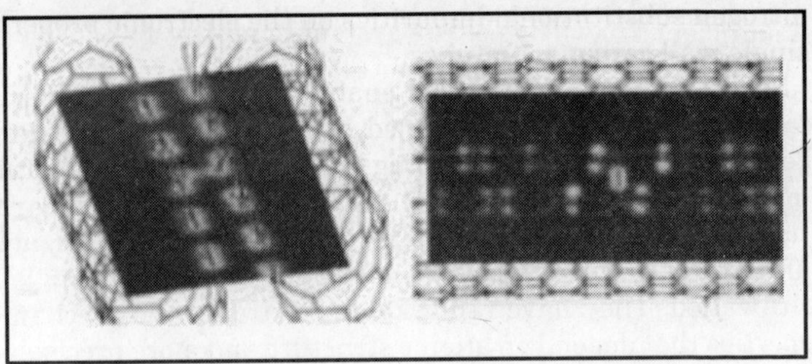

These findings open the door to the possibility of nanotube manipulation via the formation of tunnel junctions between suitably doped nanotubes. Nanotube properties could also be controlled by selective functionalization through ligand docking at the impurity sites.

CASTEP enabled researchers to treat a system of several hundred atoms, necessary in order to study the intertube covalent bond and the isolated impurity, whose electronic state decays very slowly.

In the future, researchers hope to study applications of the doped nanotubes, such as the tunnel junction or an enhanced gas sensor. This will require computing non-equilibrium electronic structures, which is at the cutting edge of current quantum mechanical modeling.

Drug Delivery

Nanotechnology is impacting on the design and development potential new drug delivery systems in many ways. In particular, researchers are studying nanoscopic carrier systems such as dendrimers and block-copolymers which 'self-assemble' to nanoscale structures that have many advantageous properties, such as increased circulation time, potential to pass the blood-brain barrier and can be functionalized for targeted delivery.

Mesoscale modeling tools are particularly useful in the study of such nanoscale large objects such as micelles, vesicles, and colloids. To obtain thermodynamic and kinetic information on such systems e.g. their phase morphology and release behavior, it is unpractical to retain all atomistic detail. Rather one wants to retain just sufficient knowledge of the system to make the information obtained from the model meaningful.

DPD and MesoDyn in MS Modeling do exactly this. Using mean-field interaction parameters obtained either from experiment or from fully atomistic systems, and coarse graining the scales (many atoms become a single bead, time steps can be made larger since these units are soft) we are able to study the whole system dynamically.

LET Small the Application of MesoDyn to Modeling Micelle Drug Carriers

Novel drug delivery systems based on molecular self assembly at the nanoscale are receiving increasing attention as a means of improving issues of drug solubility, passing the blood-brain barrier, and targeted drug delivery. Amphiphilic block-copolymers are being studied extensively for this purpose, as they spontaneously form nanoparticles known as micelles in an aqueous medium. These micelles are characterised by a hydrophobic core surrounded by a hydrophilic corona. The core region can act as a container for hydrophobic or otherwise purely soluble drugs, which in this way can be transported through the bloodstream.

Experimental investigation of the formation and structure of these nanoscale aqueous systems is very difficult, requiring complex and time-consuming preparation techniques.

Accelrys' mesoscale simulation methods MesoDyn and DPD have been developed specifically for investigating nanostructure formation in complex fluids.

Researchers at the University of Cambridge, UK, and Accelrys have used MesoDyn to simulate micelles formed from ethylene oxide—propylene oxide block copolymers in aqueous solution. They compared the simulation with direct observation of real micelles by cryo-transmission electron microscopy. Good agreement was found for different formulations, and the simulations could be used to map out the effects of varying the polymer molecular weight, aqueous concentration, and drug loading on the nanostructure.

The nanoscale structure of an aqueous solution of amphiphilic block-copolymer with haloperidol drug at 1 per cent concentration. The polymer formes micelles with a hydrophobic core and a hydrophilic corona. In the image the interface between core and corona showing that the drug mostly resides in the core of the micelles. Size scale of the image is about 35nm.

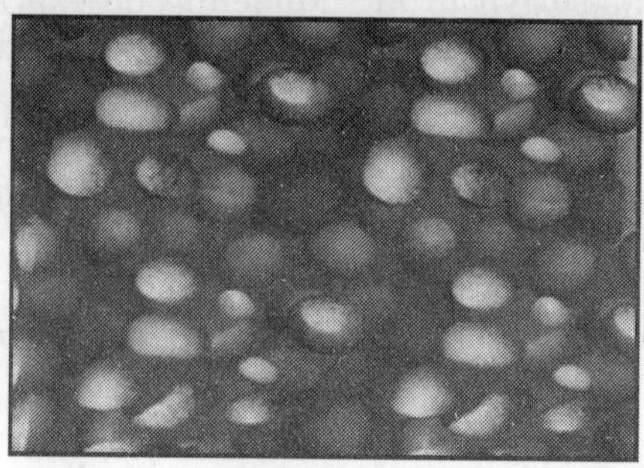

In particular, the simulations revealed that the drug tends to be located in the interface region for low loadings and become more aggregated in the core for higher loadings. The micelle shape is distorted towards a rod-like structure at higher drug loading.

Scanning Probe Microscopy Combined with Mesoscale Simulations

MS Modeling has been used to model the phase behavior of cylinder-forming block copolymers in thin films.

Using a combination of mesoscale simulations and scanning force microscopy experiments, the studies revealed that the phase behavior in such systems is dominated by surface reconstructions.

This finding will help to understand and control thin film structures in ordered fluids, including surfactant based systems.

The work demonstrates that mesoscale simulations and scanning force microscopy are a powerful complimentary approach to tackle complex nanostructured materials.

At the mesoscale (10-1000 nm), ordered fluids, for example amphiphilic block copolymers, are interesting as they show a crystal like order. At the same time they exhibit a fluid-like disorder on the molecular scale and behave on macroscopic length scales like fluids. At interfaces and in thin films further order is induced as one component usually has a lower interfacial energy than the other and tends to accumulate at the interface.

Near surfaces and in thin films a number of interesting phenomena have been observed such as the existence of wetting layers, spherical microdomains, perforated lamella, and cylinders with necks.

Despite numerous experimental and computational studies, the origins of these phenomena remain unclear.

In order to clarify matters, researchers at Bayreuth University (Germany) and Leiden University (The Netherlands) used the MesoDyn mesoscale tool to model the phase behavior of cylinder-forming polystyrene-block-polybutadien-block-polystyrene (SBS) triblock copolymer in thin films.

Robert Magerle's team determined experimentally a phase diagram for the SBS triblock copolymer in thin films using well-controlled sample preparation and scanning force microscopy. To gain further insight into the structure formation process, the phase behavior of the system was modeled using MesoDyn.

MesoDyn simulation of a film of cylinder forming triblock copolymer forming a perforated lamella surface reconstruction at the film surface.

Structural deviations from the bulk structure, so called surface reconstructions, such as wetting layer, perforated lamella, and lamella, both near surfaces and in thin films of the SBS triblock copolymer.

These findings, taken in context with previous results, support a general mechanism for the phase behavior of block copolymers at interfaces and in thin films. The structure near the surface depends on the strength of the preferential attraction of one of the components to the surface and the deformability of the bulk structure, affecting the orientation of the microdomain structure or causing surface reconstructions to form.

This work could lead to a better understanding and eventually control of thin film structures in a wide-range of ordered fluids, including surfactants and lubricants.

Image from a MesoDyn simulation of the pase separation of diblock copolymer. The surface links point at the dividing surface between the two phases. The entire volume is color-coded by the density of one of the blocks. This image is a state prior to equilibrium, and the system ultimately evolves to a lamellar phase.

5

APPLICATION OF NANOTECHNOLOGY

Many Potential Applications have Already been Idertifi ed

- (i) Consumer electronics and computing
  - ➢ Miniaturized supercomputers
  - ➢ Terabit non-volatile memory
  - ➢ Pervasive computing
  - ➢ Low voltage and high brightness displays
- ii) Chemicals and basic materials
  - ➢ Ultra-lightweight, high-strength, precision-formed materials
  - ➢ Nano-composite polymers for structural and electronic applications
  - ➢ High efficiency and novel catalysts
  - ➢ Wrinkle/Stain/H_2O resistant textiles
- (iii) Pharmaceuticals and medical products
  - ➢ New and more effective drug compounds and nearly perfectly targeted drug delivery
  - ➢ Diagnostics and sensorsfor point of care and biowarfare detection
  - ➢ DNA sizing and sequencing
  - ➢ Bioelectronics
- (iv) Energy
  - ➢ Thin film photovoltaics for cost effective solar energy
  - ➢ Cost competitive fuel cells for automotive applications
  - ➢ High capacity, rapid charge batteries

Expected Benefits from Nanotechnologies

Technology analysts highlight nanotechnologies as benefiting today or likely to benefit in the foreseeable future:

- materials sciences (esp., ceramics; more generally, lighter and stronger materials);
- cosmetics (e.g., non-ghosting sunscreen, nano-liposome-based skin care products);
- house-cleaning products (e.g., window-washing sprays);
- paints, vanishes, and other coatings;
- chemistry (e.g., tailor-made catalysts);
- information and communication technology (e.g., nano-electronics);
- biomedical applications (e.g., "lab-on-a-chip", biosensors, medical imaging, prostheses and implants, drug delivery devices);
- environmental remediation technology;
- energy capture and storage technology (e.g., solar cells, batteries, fuel cells, fuels and catalysts);
- agriculture (e.g., sensors, seed improvement);
- food (ranging from non-permeable membranes and, national legislation permitting, antibacterial powders to pathogen and contaminant sensors, environmental monitors, and remote sensing and tracking devices);
- military technology, textiles, surface finishing and lubrication agents.

(1) ENVIRONMENT

Our environment is the most important resource that we have and one that we have often taken for granted. One of the environmental concerns facing the world today is the possibility of global warming resulting from a build-up of what are called greenhouse gases, some of which occur naturally. As the sun heats the earth's surface, the earth radiates energy back into space. Atmospheric greenhouse gases trap some of this outgoing energy and retain heat. The natural greenhouse effect keeps temperatures on the earth mild enough to sustain life. However, according to the National Academy of Sciences, the Earth's surface temperature has risen by about 1 degree Fahrenheit in the past century, and this warming has accelerated during the past twenty years. There is strong evidence that most of the warming over the last 50 years is caused by the build up of greenhouse gases—mostly carbon dioxide, methane, and nitrous oxide. Why is this happening? Most

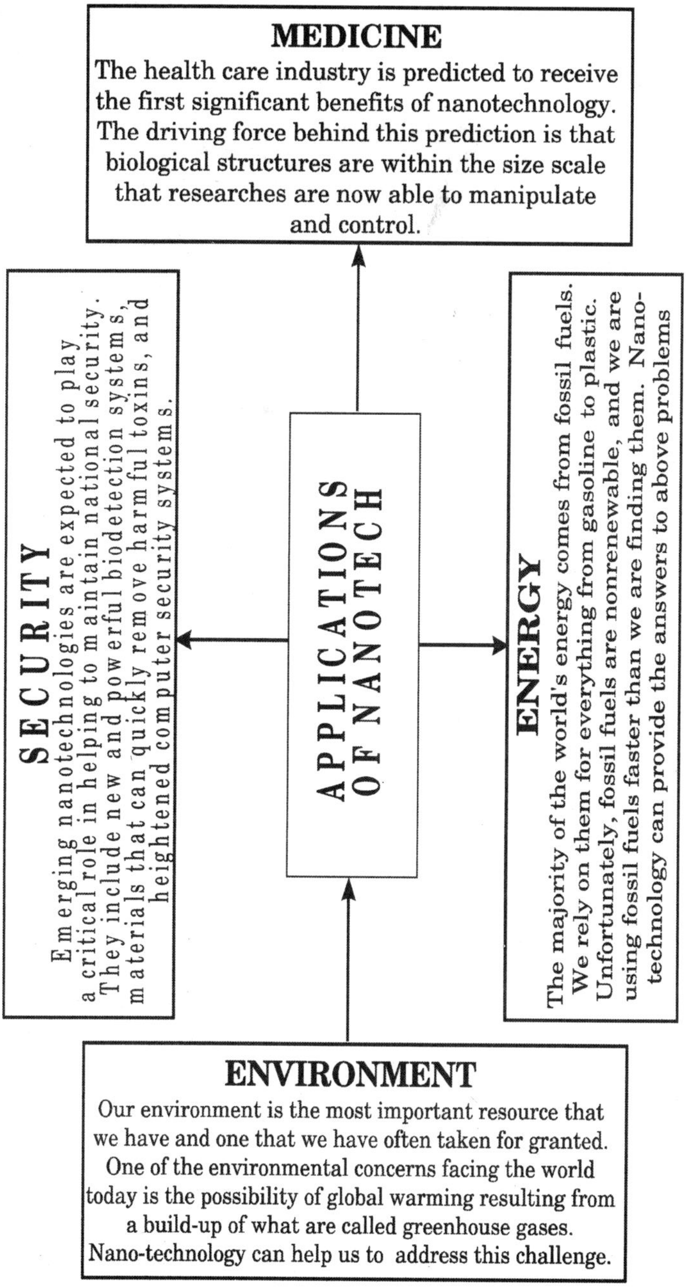

Types of Applications

MEDICINE

The health care industry is predicted to receive the first significant benefits of nanotechnology. The driving force behind this prediction is that biological structures are within the size scale that researches are now able to manipulate and control.

SECURITY

Emerging nanotechnologies are expected to play a critical role in helping to maintain national security. They include new and powerful biodetection systems, materials that can quickly remove harmful toxins, and heightened computer security systems.

APPLICATIONS OF NANOTECH

ENERGY

The majority of the world's energy comes from fossil fuels. We rely on them for everything from gasoline to plastic. Unfortunately, fossil fuels are nonrenewable, and we are using fossil fuels faster than we are finding them. Nano-technology can provide the answers to above problems

ENVIRONMENT

Our environment is the most important resource that we have and one that we have often taken for granted. One of the environmental concerns facing the world today is the possibility of global warming resulting from a build-up of what are called greenhouse gases. Nano-technology can help us to address this challenge.

scientists believe that the increased use of fossil fuels to run cars and trucks, heat/cool homes, operate industry businesses, and power factories is the primary reason.

In June 2005, the U.S. National Academy of Sciences joined similar groups from nations around the world in recognition of the potential dangers ahead. They issued a call for action to reduce greenhouse gas emissions.

Can Nanotechnology Help us Address this Challenge?

Renewable Energy

It is estimated that approximately 85% of the energy needs in the world are met by fossil fuels (coal, oil, and natural gas). These are also known as unrenewable energy sources, meaning there is a limited supply that cannot be replenished. Renewable energy sources on the other hand are constantly replenished. Most renewable energy comes from the sun either directly (i.e., solar energy) or indirectly (i.e., wind power which is controlled by the sun). Renewable energy technologies are often referred to as clean energy since they are far less harmful to the environment than conventional energy sources. Nanotechnology is offering a range of new opportunities. For example, nanotech researchers are working on the development of a solar panel/fuel cell combination. The idea behind the technology is that when the solar panel is producing energy, the fuel cell is running in reverse to collect excess energy, convert it to hydrogen, and store it. When the sun goes down and the solar panel is no longer producing energy, the fuel cell will run forward and produce energy from the hydrogen it has stored.

Remediation—Cleaning Up

During the Cold War period, the U.S. created a vast nuclear weapon research, development, and testing infrastructure. The result of this extensive activity is that approximately 7,000 sites across the country have subsurface contamination. The Department of Energy estimates that this includes groundwater equal to 4 times the U.S. daily water consumption, and enough soil to fill 17 professional football stadiums. With current technology, cleaning up this problem could take 70 years at a cost of $300 billion dollars.

Nanotechnology could provide cost-effective solutions to this and other challenging environmental cleanup problems. Researchers are currently experimenting with nanoparticles that have the ability to detoxify a wide variety of common contaminants. In one study at the University of Florida, researchers are using nanotechnology to develop composite materials that can sense mercury vapor in the air and absorb it. They have also shown that the nano-composite material will release the absorbed mercury when exposed to a heat or vacuum treatment, providing recycling opportunities.

Binding Pollutants

The problem is that modern life produces pollution—whether from exhaust fumes, chemical plants or fertilisers in the soil. However, many *'natural' pollutants also exist.* Many existing technologies fail to cope with these problems, however new nanomaterials can be manufactured which can bind to these pollutants and then be mopped up (as you would use a sponge to mop-up spilled water). For example, arsenic in groundwater is at levels above safety limits set by the World Health Organisation (WHO) in many countries, such as Bangladesh.

There is no conventional method which can reduce levels below recommended limits. However, a new system in development by Rice University uses magnetic iron oxide nanoparticles which *bind more than 99 per cent of the arsenic present,* and can then be removed from water by applying a magnetic field.

(2) ENERGY

The majority of the world's energy comes from fossil fuels— primarily coal, oil, and natural gas. All three were formed on Earth about 360 million years ago during the Carboniferous Period and long before the age of the dinosaurs. We rely on fossil fuels for much more than gasoline to power our cars. For example, tremendous amounts of oil are required to produce all plastics, all computers and high tech devices.

According to the American Chemical Society, it takes 3.5 pounds of fossil fuels to make a single 32 megabyte DRAM computer chip, and the construction of a single desktop computer consumes ten times its weight in fossil fuels. Our food is produced by high-tech, oil-powered industrial methods of agriculture, and in the US each piece of food travels about 1,500 miles before it reaches the grocery store. Pesticides are made from oil, and commercial fertilizers are made from ammonia, which is made from natural gas. Fossil fuels are needed to make many medical devices and supplies such as life-support systems, anesthesia bags, catheters, dishes, drains, gloves, heart valves, needles, syringes, and tubes. There is a limited supply of fossil fuels and they are nonrenewable. Today, we are using fossil fuels faster than we are finding them. In fact, the Oil Depletion Analysis Center (ODAC) predicts that in the near future the demand for fossil fuels will far exceed the Earth's supply.

Energy Efficiency

Researchers are exploring ways in which nanotechnology could help us accomplish the following two goals:

1. Access and use fossil fuels much more efficiently so that we can get more energy out of current reserves, and
2. Develop new ways to generate energy. One example of processes being developed to use fossil fuels more efficiently is the current research to design zeolite catalysts at the nanoscale.

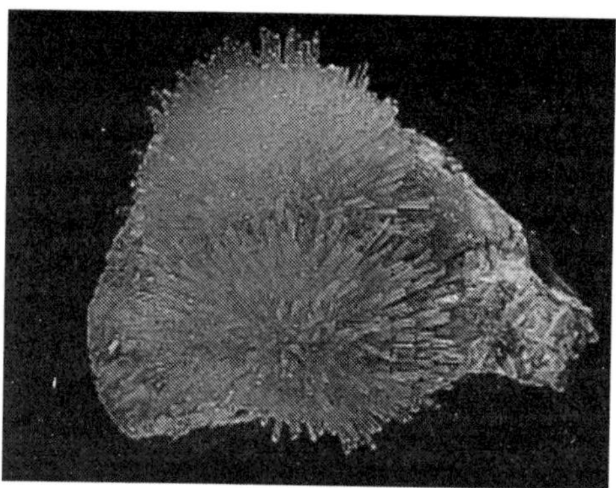

A zeolite is an inorganic porous material which works as a kind of sieve allowing some molecules to pass through while excluding or breaking down others. Zeolites can be either natural or synthetic, and new zeolites are still being discovered and invented. In 1960, Charles Plank and Edward Rosinski developed a process to use zeolites to speed up chemical reactions. Plank and Rosinski's process used zeolites to break down petroleum into gasoline more quickly and efficiently. Today, researchers are working to design zeolite catalysts at the nanoscale. By adjusting the size of the zeolite pores on the nanoscale, they can control the size and shape of molecules that can enter. In the case of gasoline production, this technique could mean that we would get more and cleaner gasoline from every barrel of oil.

New Energy Producers I

Researchers around the world are working on the development of new energy sources. What are the alternatives and what role could nanotechnology play? We already generate energy through hydropower (i.e., water). Damns have been built on most of the major waterways. Energy generated by wind turbines could help to lessen some of our dependence on fossil fuels. Although wind turbines are in use today, their energy output is far less than what is needed. Harnessing the sun's rays to make solar power is another alternative, but with current technology, solar panels could only make a limited impact.

Of these three methods, solar energy holds the most promise. Right now we need silicon wafers to make solar panels, and silicon wafers, like computer chips, require a large amount of fossil fuels for production. However, researchers like Paul Alivisatos at the University of California, Berkeley hope to use nanotechnology to develop nano solar cells that would be energy-intensive and far less expensive to make. Researchers at the University of Toronto are using nanotechnology to develop solar panels capable of harnessing not only the visible light from the sun, but the infrared spectrum as well, thus doubling the energy output. What's more, these new solar cells could be sprayed on surfaces like paint, making them highly portable. Researchers at Rice University want to take solar energy research even further. They hope to someday build a solar power station in space capable of catching the solar energy that bypasses the Earth every day and providing about nine times the efficiency of solar cells on Earth. In these and other energy-producing advances, nanotechnology will play a critical role.

New Energy Producers II

Another area where nanotechnology is expected to play a major role is in the development of new fuel cells.

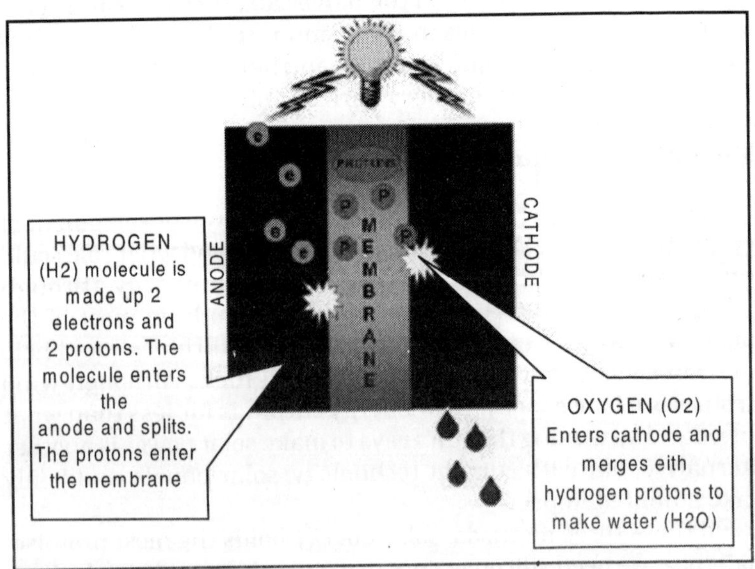

The first fuel cell (or "gas battery") was invented by William Robert Grove in 1839, only 39 years after Alessandro Volta invented the battery (the voltaic cell). Unfortunately, the materials that Grove used to make his fuel cell were unstable and the technology was unusable. One hundred and twenty years later, NASA revived fuel cell technology with new materials and used it on manned space flights. These new fuel cells were quiet, reliable, and clean, and produced water as a by-product. It was an ideal scenario. The fuel cells produced both power and drinking water for the astronauts. Today, there are a number of different types of fuels cells under development. But there are still some challenges with this technology—challenges that nanotechnology may be able to address.

Nanotechnology for Energy—Insulation

Humankind has used natural materials for insulation throughout its history. While traditional materials work well in settings where bulk application is possible, such as wall cavities, they are **not** **suit**able for applications such as glazing where significant amounts of heat can be lost or gained.

Nanotechnology offers enhanced insulation allowing thinner coatings or fillings to prevent heat loss or gain which would not be possible with conventional materials

Aerogels and Nanofoams

Aerogels are a form of insulation based on nanotechnology. They can be anything from 2-8 times more effective than traditional insulation, thereby providing extensive heat and energy savings, and consequently benefitting the environment. The reason they are so effective is that they are made up of very little solid material; approximately 96 per cent of their volume is air. Most aerogels are made from silicon or carbon.

Aerogels have a variety of uses and applications, and should not be thought of simply for the home. They can provide improvements wherever superior thermal, fire and acoustic barriers areneeded, and are usually smaller and lighter than conventional insulation methods. Aerogels can currently be found in use in oil pipelines, natural gas and military ships, building and construction, boots and jackets and much more

Nanotechnology for Energy—Thermoelectricity

Thermoelectricity is the conversion of heat to electricity and vice versa. It relies on connecting two different electrically conducting materials, which conduct heat at different rates, at two junctions (thermocouples) in a closed loop. Applying heat at one thermocouple while keeping the other cool generates an electric current within the loop. Combining many such loops produces a thermopile which can be used to power devices e.g. radios and clocks. In contrast, if electricity is passed through the loop then the thermocouples will heat or cool (which can be used for heaters or fridges). Both metals and semiconductors can be used to generate this effect.

Thermoelectric generators have advantages over those using conventional materials as they have a smaller size, and no mechanical parts that can fail or deteriorate over time.

Higher Power

Thermoelectric devices have been used for almost a century, however their efficiencies are very poor. For example the solid state fridges that have appeared on the market in recent years (which are the opposite of a thermoelectric generator and use electricity to cool contents, or if the polarity is reversed to warm contents) have an efficiency about 25% that of regular fridges.

It is only relatively recently that the reasons behind this lack of efficiency have been understood. Certain nanomaterials have been developed which have efficiencies 3 or 4 times greater than the previous best semiconductors. The reason is in the nanoscale structure of these materials which helps slow down heat transfer , while still allowing electrons to move freely. By combining several of these thermocouple of nanostructured materials together, thermopiles capable of generating reasonable quantities of electricity can be generated.

In the future it is envisaged that such thermopiles could convert waste heat from vehicle engines and exhausts to electricity, removing the need for alternators to power electrical components and recharge the battery. They also have applications in running sensors and small devices such as watches from a person's body heat.

New Applications

Thermoelectric devices have other useful applications apart from generating electricity. In fact the first commercial application of nanostructured thermoelectric materials is in the computing industry for cooling microprocessors. The market for thermal management in the electronics industry is huge (about 3.3 billion USD, and is expected to grow approximately 12% per year to 6 billion USD by 2008). Conventionally microprocessors were kept cool (within their operating temperatures) by mechanical fans. However the increased numbers and densities of transistors on the more powerful computer chips has made this almost impossible to achieve using fans.

Nanostructured thermoelectric materials on the otherhand can be highly effective at cooling such chips just by applying as microdots. In this case one side of the microdot absorbs heat, which is then emitted from thermocouples at the other side.

Nanotechnology for Energy—Portable Power

Portable devices are commonplace in modern society, from consumer products such as mobile phones and MP3 players, to medical diagnostic equipment, to environmental and chemical sensor equipment. In all cases increased functionality, smaller size and longer operating times are desired. This can include improvements to existing systems such as rechargeable batteries to supply more power and energy. It can also include new technologies such as fuel cells which generate electricity from hydrogen (see separate section).

Increases in the energy and power output of portable supplies can not only affect portable devices, but at the other end of the scale could offer *breakthroughs for all-electric vehicles*, for off-grid power systems and for high-technology applications such as aeronautics. The global battery market is worth some *30 billion USD* and is increasing dramatically

Higher Power—Supercapacitors

Capacitors store energy in the form of electrical charge, unlike batteries which store energy chemically. These have the benefit of not losing power during storage, and being able to deliver power and be recharged quickly. However they have limited reserves and are really only good for short bursts. Supercapacitors combine the advantages of both capacitors and rechargeable batteries by being able to store large amounts of energy. For example they are now being used in hybrid vehicles (such as the Prius pictured right) for "regenerative braking" which stores energy that would normally be lost as heat during braking. This stored energy can then be used by the electric motor when the vehicle starts moving again.

Nanotechnology has driven the development of supercapacitors through the production of novel nanomaterials with increased surface area. Such materials can accommodate much more charge than conventional materials, thus increasing energy density and power output many fold.

Portable Power—Rechargeable Batteries

The main *issues with current rechargeable batteries* are: lack of power output, rate at which batteries can be recharged and discharged, loss of charge over time (even when not in use), and lifespan (number of times a battery can be recharged). Nanotechnology can offer solutions to each of these. For example, the latest line of Sony camcorder batteries uses nanotechnology to increase lifespan and also decrease charge-time.

The most efficient rechargeable batteries use lithium. The total energy that a battery can hold is proportional to the amount of lithium that it contains. Lithium, in the form of positively charged ions, migrates between the two electrodes, moving to the anode (negative electrode) during the course of charging a battery, and the cathode (positive electrode) when the battery is in use (i.e. discharging). Higher energy batteries can be achieved by incorporating more lithium, however this puts strain on the electrodes which have to cope with large increases in volume due to the migration of lithium during the charging and discharging cycles (this can lead to fracturing of electrodes). Higher power is achieved by increasing the speed at which the lithium ions can migrate between electrodes.

Nanocomposites

Polymer/inorganic nanocomposites are composed of two or more physically distinct components with one or more average dimensions smaller than 100 nanometers. From the structural point of view, the role of inorganic filler, usually as particles or fibers, is to provide intrinsic strength and stiffness while the polymer matrix can adhere to and bind the inorganic component so that forces applied to the composite are transmitted evenly to the filler.

The use of nanomaterials has allowed both high energy and high power batteries to be achieved. Nanocomposites of lithium and electrode materials keep the bulk lithium in small particles which are easily broken up during the discharge process, and the porosity of the electrodes let the ions exit and enter much more rapidly. Such nanocomposites also help reduce self-discharge (i.e. loss of power during storage). Advances in electrolytes (that allow the passage of the lithium ions between electrodes) also enhance these effects.

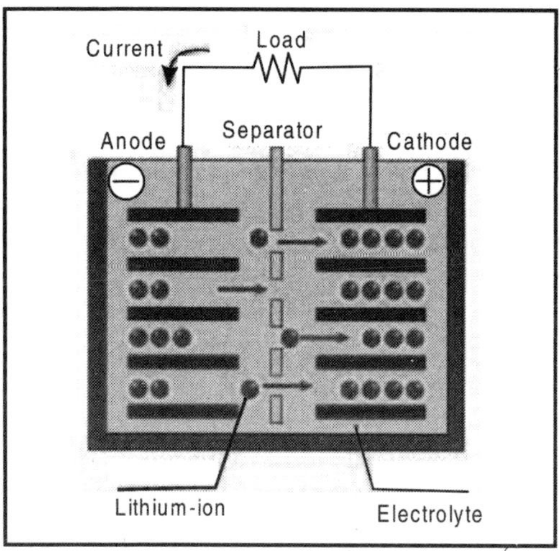

Nanotechnology for Energy—Solar

The efficiency of a solar cell depends on how much of the available spectrum of light it can absorb (the rest is simply reflected or lost as heat), and how effectively it converts this energy into electricity.

The problems with current methods are the expense of material and relatively low efficiency. This means that solar energy is several times more expensive than energy derived from burning fossil fuels.

New nanomaterials have a higher efficiency than silicon used in existing solar panels. Nanotechnology can also offer new applications such as flexible panels.

Solar—Higher Power

Silicon-based solar cells based on existing microtechnologies have a maximum efficiency of about 25% (in the laboratory) and about 14% from those that can be bought from stores. The low efficiency is down to two things: silicon absorbs light within a very defined spectrum to convert into electrical current (this is known as the bandgap, the rest of the light is either not absorbed or radiated as heat) and secondly, some of the absorbed energy is lost through poor conductivity within the solar cell. New nanomaterials and structuring solar cells at the nanoscale can help overcome both of these obstacles.

Semiconductors other than silicon can be used for solar cells (such as gallium, indium, and germanium). Each semiconductor (or mixed compound of semiconductors) has a different bandgap; so if combinations of these are applied one on top of the other in thinfilms (a few tens of nanometres thick) then each layer can absorb different parts of the spectrum of light, thus increasing the overall amount of energy absorbed. Known as multi-junction solar cells, these have achieved an efficiency of 35%. Another key aspect of these types of solar cell is that the crystal structure of each layer must be precisely matched to ensure efficient electrical transfer. Note that such semiconductors can be used "in reverse" to produce different colours of light in light emitting diodes (LEDs).

In the future Quantum Dots are predicted to provide the highest efficiency solar cells at around 85%. This is because Quantum Dots can be produced in different sizes and chemical composition to capture all available light.

Nanotechnology for Energy—Hydrogen

Most of our energy needs are met by combustion—of coal, gas and oil. However this is not the most efficient way to extract energy from fuels. Fuel cells do so chemically by reacting hydrogen and oxygen gases together to make water, heat and electrical charge.

The mechanism of generating energy varies between fuel cells, however in each case a charged ion (which is usually hydrogen or oxygen), is produced at one electrode and migrates to the other through a selectively permeable membrane, while the electrons produced as a result of this travel through an external circuit, powering devices.

Fuel cells fall into two groups: those with a high operating temperature (above 200°C) which are suitable for supplying power to whole buildings (such as solid oxide and molten carbonate fuel cells) and lower temperature units which are suitable for powering vehicles and mobile devices (such as direct methanol and polymer exchange membrane fuel cells). The high temperature units can use natural gas to form hydrogen, the lower temperature units, with the exception of direct methanol fuel cells, require hydrogen gas itself.

Hydrogen at the moment is largely produced from natural gas, however in the future it is hoped that renewable energy could be used to supply the necessary power to split (electrolyse) water into hydrogen and oxygen (i.e. the reverse of the reaction in a fuel cell).

Nanotechnology can offer solutions to material costs, fuel cell efficiency, and storage of hydrogen feedstock.

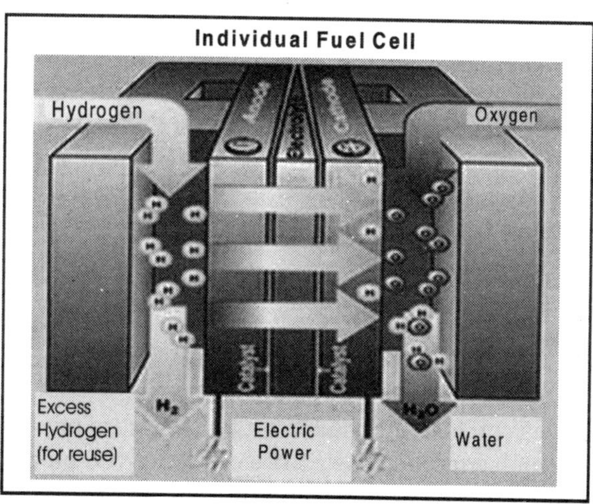

Storage

Hydrogen has the potential to provide energy without burdening the environment (if produced from renewable sources such as electrolysis of water). Hydrogen is also the favourite to replace liquid fossil fuels for powering vehicles. However the key issues will be transporting and storing it safely and in a form that can provide equivalent amounts of energy to current fuels.

It is estimated that an average car would require 1.2 kg of hydrogen to travel 100 km (which is equal to about 13,500 litres of hydrogen gas). Storing hydrogen as a liquid in pressurised containers is also not practical, because of the low energy quantity per litre and the need for improved barrier systems to line the tanks and prevent hydrogen gas leakage. However transporting hydrogen as a liquid has merit for refuelling stations (this would use the same infrastructure as exists for petrol, diesel and liquid petroleum gas).

For onboard vehicle storage the only practical format is to supply hydrogen as a solid compound, such as a metal hydride. Many nanomaterials are now being tested and developed to fulfil this role including carbon nanotubes and metal organic frameworks (MOFs—pictured), which have the largest surface area of any manufactured material (1 gramme has a surface area equivalent of a football pitch). The importance of this is that a higher surface area allows a larger amount of hydrogen to be stored. The issues that still need to be addressed are improving the amount of hydrogen that can be bound and released by the compound, and ensuring this is achieved quickly (into the storage compound at refuelling stations, and from it when needed by the fuel cell).

More Efficient Use

Hydrogen fuel cell efficiency depends on efficiently forming the charged ion, and allowing this to selectively travel from one electrode to the other while the electrons are forced to pass through the external circuit.

Electrodes can be enhanced through the use of nanostructured materials which increase the surface area of the electrode. This increases the rate at which the charged ion is formed at one electrode, and the hydrogen and oxygen combined at the other to produce water.

Membranes and electrolytes are in some cases combined in one material, in others they are a "sandwich" with membrane on either side of the electrolyte. Controlling the porosity and chemical functionality of the membrane (by using composites of different materials) can enhance the transfer of the charged ion from one electrode to the other, while preventing electron flow, and prevent contamination with other reactants or products. Manufacturing the membrane from different materials can allow it to operate more efficiently at different temperatures, or increase its lifespan (particularly true of molten carbonate fuel cells which operate at high temperatures.

(3) MEDICINE

The health care industry is predicted to receive the first significant benefits of nanotechnology. The driving force behind this prediction is that biological structures are within the size scale that researchers are now able to manipulate and control. Investigators are looking to nanotechnology to develop highly sensitive disease detectors, drug delivery systems that only target the disease and not the surrounding healthy tissue, and nanoscale building blocks that help repair skin, cartilage, and/or bone.

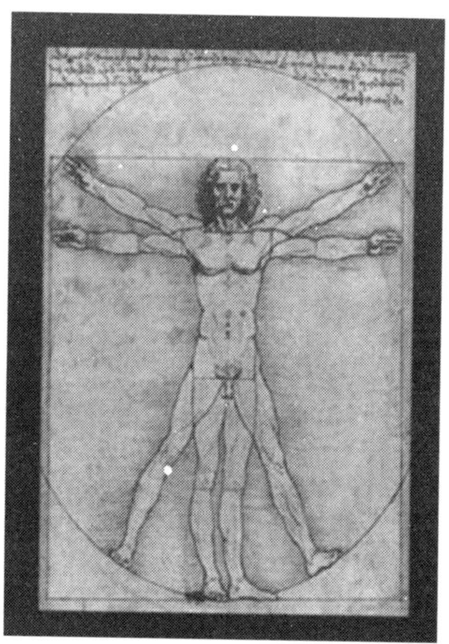

Nanotechnology has the potential to revolutionise healthcare for the next generation. There are three key areas in which it could do this: Diagnosis, Prevention and Treatment.

The vision of the future is of harnessing the qualities of nanotechnology to eventually provide healthcare which operates purely from a preventative state, identifying and stopping potential sources of disease/illness in the body before they even get started.

Prevention

The majority of healthcare is reactive rather than preventative. This is often some time after the initial infection or trauma, which means that tissue damage and suffering has already occurred. In some cases this damage can be irreparable, in others permanent reminders remain (such as loss of function of that body part or scarring).

Therefore, possibly the most important aspect of nanomedicine in the future will be its potential to prevent illness, rather than simply treating it. Nanotechnology will contribute to this through more effective monitoring of individuals' health (allowing diseases to be caught in their infancy) and more sterile hospital environments (limiting the opportunity for bacteria, viruses and other microbes to cause secondary disease). Understanding the genetic make-up of the patient will also allow the doctor to prescribe personalised medicine.

Antimicrobial Coatings

Antimicrobial coatings can give major benefits in areas of healthcare, by helping to minimise the persistence and spread of microbes such as viruses, bacteria and fungi. Such coatings are seen as an addition to, and not a replacement for, the procedures in place for decontaminating and sterilising surgical equipment and operating theatre surfaces (such as disinfectants, autoclaving etc). Coatings can help minimise the ability of microbes to bind and start growing on surfaces that are exposed to patient body fluids during normal operating procedures. This can be achieved passively (through "non-stick" coatings, as described in the implants section) and actively (through the inclusion of coatings such as silver and titanium dioxide nanoparticles that can kill microbes directly). Such coatings can also help minimise the accidental spread of disease from surface to surface to patient.

Silver nanoparticles are also making their way into the home. In recent years, major manufacturers such as Daewoo and LG have been coating the inside of their fridges with them. This anti-bacterial surface helps maintain a healthy, clean environment, and means the contents of the fridge stays fresh longer.

Implants

Implants can perform a number of different functions: from joint replacement, to stents which keep arteries open, to active implants such as cardiac pacemakers and cochlear implants (to restore hearing loss). In all cases such implants must interact closely with biological tissues. One of the key issues is encouraging the patient's cells to stick to the implant where required (e.g. for bone grafts) thus assisting in the repair of a damaged tissue, while in other cases ensuring that no cell or biological material sticks (such as the interior of arterial stents).

Over the last few years it has become clear that both chemical signals and physical characteristics of materials effect the ability of cells and biomiolecules to stick. For example, nanoscale bumps and grooves can increase cell adhesion, whereas completely smooth surfaces are very poor at allowing cells to stick. Decorating surfaces with similar molecules to those that are found in tissues also increases "stickiness". The applications are seen for example in titanium bone implants that have a coating of nanostructured titanium dioxide which improves integration in the bone, and diamond-like coating of stents and catheters which are smooth and have marked decrease adhesion of blood proteins and cells.

Nanotechnology can also take the potential for the rehabilitation of the weak and old to a totally new level. Some of the devices presently being researched include retinal implants. Many people suffer of degeneration of the retina with old age. One solution is to use a photosensor array which will detect incoming light and connect it to a signal processor. A signal can then be transmitted to an implanted receiver at the retinal interface which is connected via microcontact to the retinal nerve.

Ceramic Nanocomposites in Artificial Joints

The European Commission reports that ceramic nanocomposites could solve the problem of fracture failures in artificial joint implants. This would extend patient mobility and eliminate the high cost of reparative surgery.

Bone

Nanotechnolgy is being used in teeth and bone replacements copying the way nature itself lays down minerals. This process is called biomimicry. Biomimicry is already the basis of new tough and light materials for bullet proof vests and other defence applications.

The use of nanopatterned polymers could eliminate the long recovery times, scarring and infection associated with and bone grafts. Researchers hope to use this technique to grow adult stem cells that will turn into bone.

Artificial Skin

The application of nanotechnology will result in artificial skin, reconstructed tissue and wound treatments that are better, durable and more acceptable. This will help in the regeneration of tissues and even whole organs will be able to be grown to replace organs that have failed due to disease or old age.

Tissue Engineering

Tissue engineering at the nanoscale level is leading to the development of viable substitutes which can restore, maintain or improve the function of human tissues.

Cell Engineering

Chemistry, nanotechnology and materials science are all important tools used by cell engineers to study and control how cells behave.

Cells react on the micro and nano scale to the shape and chemistry of their surrounding environment. Nanoscale grooves no wider then the sells themselves can act as templates causing cells to line up. Once cells can be induced to organise themselves, even more possibilities open up, ranging from wound repair through to the future vision of growing whole organs.

Artery replacement is also possible using structures made form natural polymer to act as a scaffold around a patients natural artery, which given the appropriate stimulus, can be encouraged to regenerate. These scaffolds being biodegradable melt away after the regenerated artery is in place.

Nanocapsules

Nanocapsules encapsulate a drug's active component in a relatively inert 'nanocapsule', which binds and opens in response to at arget tissue site. A self-assembling nanoscale polymer carries anti-cancer drugs across the blood-brain barrier. This then targets the affected tissue only thus producing less side effects.

Nano-particles

Magnetically coupled inorganic manufactured nano-particles are drugs being drawn to a drug target via an externally applied magnet. This places nanomagnetic particles into the tumours. Under the influence of a magnetic field, these particles heat up and dissolve the tumour cells within the body. This is seen as a more effective treatment of cancer.

Nanospheres

A Nanosphere incorporating therapeutic agents allows better penetration of the particles inside the body. Its size allows delivery via intravenous injection and therefore they can be used for intramuscular or subcutaneous applications. This minimises the irritant reactions at the injection site.

Dendrimers

Dendrimers go through the vascular pores and into tissue more efficiently than larger carriers. They have a high drug-carrying capacity that can release a heavy payload (desirable with cancer drugs) without damaging tissue.

Nanoporous Membranes

Nanoporous membranes act as tiny turnstiles for releasing drugs.

By making the nanopores only slightly larger then the molecules of drugs, they can control the rate of diffusion of the molecules keeping it constant, regardless of the amount of drug remaining inside a capsule. Nanoporous membranes can also be used in implantable devices for the treatment of chronic disease.

Nanoparticulates Fight Against TB, Diabetes

Nanoparticulates are used for inhalation technology. Polymeric 'Trojan' nanoparticles are designed to efficiently deliver to the lungs particles possessing dimensions and mass too small to otherwise deposit effectively. This allows sustained drug action and release throughout the lungs. This long-action therapy through the process of inhalation is important for fighting diseases such as tuberculosis and diabetes.

Buckyballs Effective in Fighting AIDS

Buckyballs are a stable platform that enhances a drug's effectiveness by allowing it to more readily attach itself to its intended molecular target. The buckyball is readily absorbed into the blood stream and has the ability to cross the blood-brain barrier. Buckyballs are seen to particularly effective in fighting HIV.

Filters

One of the most important means of preventing disease is preventing exposure to pathogenic microbes. As well as providing sterile surfaces, this can take the form of filtration of air and liquids that a patient is exposed to during treatment. The trouble is that many viruses are smaller than the pores of these filters and so can penetrate them, making them useless.

New filters have nanoscale pores that are able to remove even the smallest of viruses. Inclusion of active materials, such as silver nanoparticles or titanium dioxide nanoparticles and UV light sources, can enhance this effect by killing the trapped viruses, bacteria and fungi. Such systems are already being employed in the fight against SARS, to prevent the spread of the virus from infected patients to medical staff.

Monitoring

Monitoring of a patient's health status is important not only for those recovering from operations and treatment, but also for the routine check-up of healthy individuals. Point-of-care (POC) devices offer an unprecedented degree of flexibility through the measurement of many different physiological factors such as blood pressure, blood chemistry (e.g. levels of sugars, hormones, antibodies in the blood), heart rate, and body temperature at the patient's location without the need to send samples off to the lab.

For more complicated tests, POC devices can incorporate Lab-On-A-Chip devices (pictured left) which allow tens or hundreds of different biomolecules to be measured rapidly. By measuring quickly and at the patient's location, doctors avoid the risk of losing samples, waiting days for results to come back from the lab, and misdiagnosing due to samples being stored or treated incorrectly. This allows the correct treatment to be given quickly and gives a portability that allows diagnosis of disease in remote areas (such as HIV infection in developing countries).

In the future such devices may be linked wirelessly to a computer in the doctor's office, allowing patients to monitor themselves from the comfort of their own home and only attend the doctor if a change in treatment is required.

Treatment

At present, treatment of illness tends to use quite old-fashioned, well-established methods. These methods are obviously not always entirely successful, with general solutions and treatments often being applied to very specific problems.

Furthermore, current healthcare can often cause additional problems such as rejection or a bad reaction to a transplant.

Nanotechnology can help to address some of these issues, in a number of ways. It will be possible to provide personalised medicine designed for individual patients and their particular illness.

The danger of implant rejection can be minimised through 'body-friendly' nano-coatings, whilst there are also new techniques being developed to allow drugs to be more accurately targeted.

(4) SECURITY

National Security

Since the attacks on the World Trade Center in 2001, the fundamental understanding of national security has changed. In October 2001, the Department of Homeland Security was established. Twenty-two existing agencies are now under this one organization. Their assignment is to protect the US from future attacks and ensure a timely response to any future emergencies.

Emerging nanotechnologies are expected to play a critical role in helping to maintain national security. They include new and powerful biodetection schemes that can analyze a potential bioterrorism threat at the point-of-care, materials that can detoxify an area or human exposed to a set of toxins, and novel ways of encoding structures that can be used to secure computer systems.

Sensors

As the battlefield moves from military targets to civilian, biological and chemical weaponry may play an increasingly dominant role. Protection from these threats depends on the ability to detect, respond, and control biological and chemical threats before they can harm the body.

Nanosensors' ability to detect at the molecular or even atomic level is critical. While in the realm of medicine, biosensors can detect the onset of disease; in the area of national security, they could be used to detect radioactive materials or toxins like anthrax.

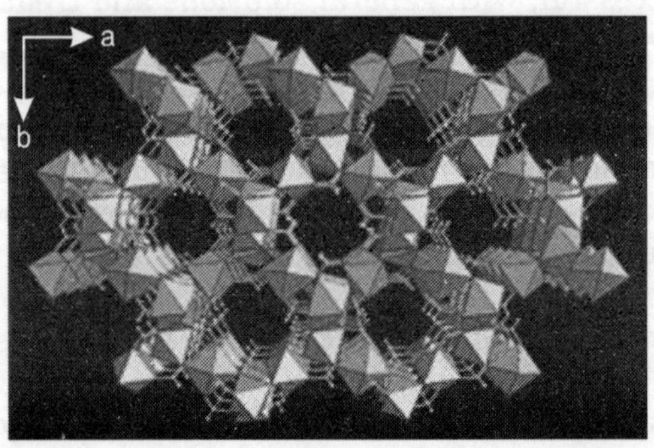

Another concern is the vulnerability of public works and public buildings. The scale of nanotechnology could mean the development of sensors embedded in clothing or painted on the side of a building. The high sensitivity of nanosensors also means that large public works, like a water system, could be routinely tested and even extremely small amounts of contaminants would be detected.

Communications

Another area of great importance to national security is that of protecting our information systems from attack. Computers and networks are the foundation of major sectors of our economy like our financial institutions and electric power sectors.

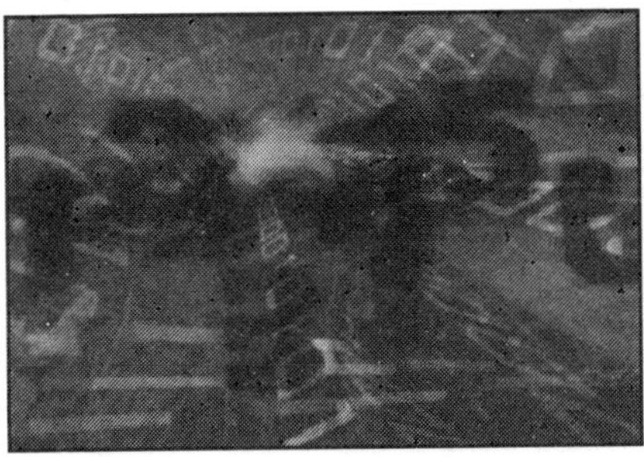

Researchers working in the areas of nanoelectronics and nanocomputing hope to integrate transistor-like nanoscale devices into system architecture, to provide substantial advantages over current technologies. They are also working on the creation of powerful "grid protocols" that could make the world wide web obsolete, and something called quantum cryptography that could provide the type of electronic security systems that are impossible to crack.

Protecting Our Troop

Another area of great importance to national security is that of protecting our troops. Many of the applications mentioned earlier are potentially applicable to the protection of the soldier in the field.

In 2003, the United States Army committed $50 million dollars to establish an Institute for Soldier Nanotechnologies (ISN) at Massachusetts Institute for Technology.

The goal of the ISN is to use nanotechnology to create uniforms and gear that will protect and heal the foot soldiers of the future. Although not the only research effort in this area, the ISN is the largest.

Other Applications

Tennis balls

Nanotech has already enhanced the life of tennis balls. The inner core of this ball is coated with a barrier to maintain the air pressure and bounce at least two times longer, increasing the playable time of the ball.

Using a patented material and process, the inner core of the tennis ball is coated with a butyl-based barrier called Air D-Fense™ by InMat™ LLC, which restricts airflow from escaping the core. This new technology utilizing a coated inner core inhibits air permeation by 200%. Other balls left out of the can become unplayable after two weeks or less. Double Core™ is playable beyond four weeks.

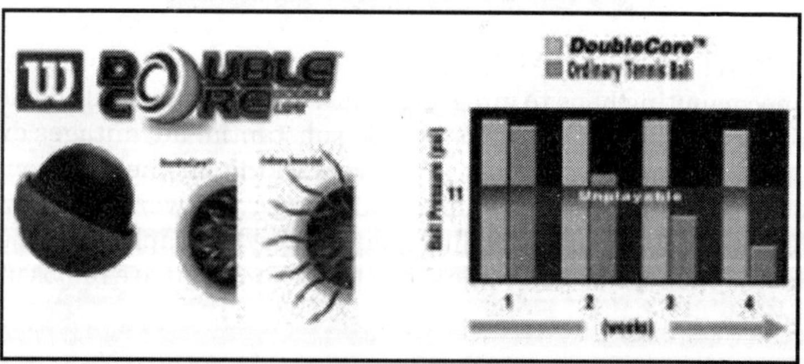

- Gas barriers are used in a wide variety of applications including soft drinks, food packaging, automobile and other tires
- Improved barriers can extend life, and improve product quality/performance

Automobile

In tyres, the new technology helps reduce weight, improve fuel efficiency, improve pressure retention and reduce recycling and incineration costs.

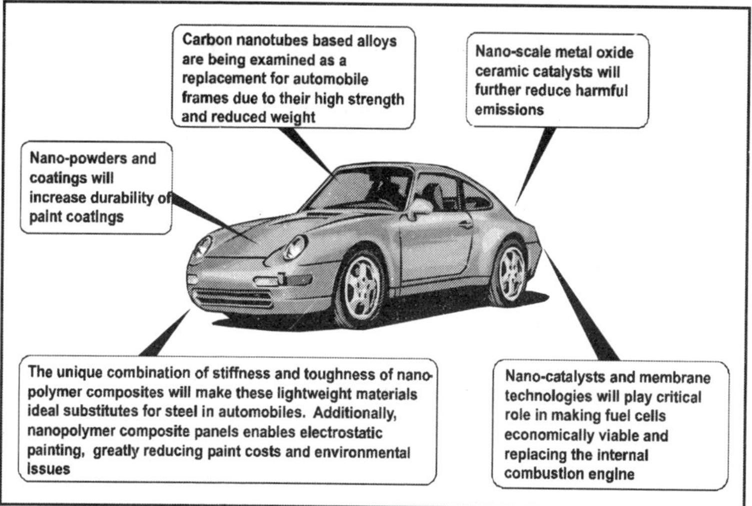

Cosmetics

In order to get complete sunblock protection, lifeguards and others apply zinc oxide. This substance has a white color. By creating nanoparticles in the 30-60 nanometer range, Nanophase Technology has created a clear formulation which blocks all UV rays, yet lets all wavelengths of visible light through

Optical barriers have use in a variety of applications:

- Sunblocks
- UV protectants (glasses, windows)
- Photovoltaics

Agricultural Productivity Enrichment

(a) Nanoporous zeolites for slow release and efficient dosage of water and fertilizers for plants and of nutrients and drugs for livestock

   (b) Nanocapsules for herbicide delivery
   (c) Nanosensors for soil quality and for plant health monitoring
   (d) Nanomagnets for removal of soil contaminants

Water Treatment and Remediation

   (a) Nanomembranes for water purification, desalination and detoxification
   (b) Nanosensors for the detection of contaminants and pathogens
   (c) Nanoporous zeolites, nanoporous polymers and attapulgite clays for water purification
   (d) Magnetic nanoparticles for water treatment and remediation
   (e) TiO_2 nanoparticles for the catalytic degradation of water pollutants

Disease Diagnosis and Screening

   (a) Nanoliter systems (Lab-on-a-chip)
   (b) Nanosensor arrays based on carbon nanotubes
   (c) Quantum dots for disease diagnosis
   (d) Magnetic nanoparticles as nanosensors
   (e) Antibody-dendrimer conjugates for diagnosis of HIV-1 and cancer
   (f) Nanowire and nanobelt nanosensors for disease diagnosis
   (g) Nanoparticles as medical image enhancers

Food Processing and Storage

   (a) Nanocomposites for plastic .lm coatings used in food packaging
   (b) Antimicrobial nanoemulsions for applications used in decontamination of food equipment or packaging
   (c) Nanotechnology-based antigen detecting biosensors for identification of pathogen contamination

Air Pollution and Remediation

   (a) TiO_2 nanoparticle-based photocatalytic degradation of air pollutants in self-cleaning systems

   (b) Nanocatalysts for more efficient, cheaper and better-controlled catalytic converters

   (c) Nanosensors for detection of toxic materials and leaks

   (d) Gas separation nanodevices

Vector and Pest Detection and Control

   (a) Nanosensors for pest detection.

   (b) Nanoparticles for new pesticides, insecticides and insect repellents

Constructions

Construction is an area in which nanotechnology can make a big difference. Indeed, some applications are already on the market. For example, in the environment section we look at new smart materials incorporating titanium dioxide nanoparticles that have given rise to self-cleaning windows, buildings and even roads (7000 square metres of road in Milan have been coated with these smart materials, which resulted in a 60% reduction in nitrogen dioxide levels).

Other nanomaterials are being incorporated in new construction materials to improve mechanical strength, durability and insulation, while at the same time decreasing the weight compared to traditional materials. For example, ceramic nanoparticles are being used in cement to increase their robustness.

Ambient sensor systems will be increasingly used in construction, to monitor a building's environment including any mechanical stresses it is placed under

Ambient Sensor Systems

Sensing is about detecting structural stresses on the building, wear and tear etc, and alerting people to weaknesses or changes the building may be undergoing.

Sensor use in buildings can also provide environmental monitoring, or even temperature control whereby it can be linked to the heating/air-conditioning system to choose an optimal setting, based on the data the sensor picks up.

Nanotechnology can also help to provide a systems-wide approach in detailing how the 'building is feeling'. All sensor and monitoring data will be communicated to a central node where the information can then be disseminated and acted upon.

The electronics involved will be smaller, but equally as (or even more) powerful. They will be more energy efficient, and could even include small solar cells to effectively allow them to power themselves. As a result, there is the potential to make the systems maintenance-free and long lasting.

Advanced Materials

Construction materials are used in many different situations and as a result have to cope with many different environmental conditions, as well as fulfilling their primary role of strength and support. The properties of materials can often be enhanced through combination, e.g. steel reinforced concrete is stronger and more resistant to deformation stresses than concrete alone. However, such reinforced concrete is at the mercy of the elements. Repeated hot and cold, and wet and dry cycles eventually cause the concrete to crumble, exposing the steel which then corrodes.

New composite materials based on advances in nanotechnology can address these problems. The addition of nanotubes and ceramic nanoparticles to materials such as concrete and polymers can increase their load-bearing abilities, increase their resistance to environmental conditions (such as extremes of hot and cold, or corrosive salt conditions, or immersion in water), and increase their ability to cope with other stresses such as collision or movement (e.g. earthquakes) without shearing or breaking. These advantages are realised through improved binding of the bulk material (e.g. concrete or polymer) leading to a highly ordered, internal nanostructuring of the composite. This replaces the need for heavy support materials (such as steel) and also eliminates the possibility of future weakening due to corrosion.

New composites can also enhance the thermal insulation properties of construction materials (see the insulation section in Energy), as well as improving fire retardancy. In this respect there is a lot of interest in using nanomaterials in combination with wood (which is still a major construction material for housing).

Not only are such materials stronger and more durable than traditional ones, they are also lighter. This means that construction using such components is quicker and does not require heavy lifting machinery. Already several bridges around the world have been repaired or constructed using composite materials based on microtechnology. Nanotechnology enabled adavances are expected to further improve on the mechanical properties of such composites

New composite materials based on advances in nanotechnology can address these problems. The addition of nanotubes and ceramic nanoparticles to materials such as concrete and polymers can increase their load-bearing abilities, increase their resistance to environmental conditions (such as extremes of hot and cold, or corrosive salt conditions, or immersion in water), and increase their ability to cope with other stresses such as collision or movement (e.g. earthquakes) without shearing or breaking. These advantages are realised through improved binding of the bulk material (e.g. concrete or polymer) leading to a highly ordered, internal nanostructuring of the composite. This replaces the need for heavy support materials (such as steel) and also eliminates the possibility of future weakening due to corrosion.

New composites can also enhance the thermal insulation properties of construction materials (see the insulation section in Energy), as well as improving fire retardancy. In this respect there is a lot of interest in using nanomaterials in combination with wood (which is still a major construction material for housing).

Not only are such materials stronger and more durable than traditional ones, they are also lighter. This means that construction using such components is quicker and does not require heavy lifting machinery. Already several bridges around the world have been repaired or constructed using composite materials based on microtechnology. Nanotechnology enabled adavances are expected to further improve on the mechanical properties of such composites.

Breaking Down

Environmental remediation can benefit from the use of nanotechnology. At the forefront of this is the use of titanium dioxide nanoparticles. These function as catalysts which assist in the breakdown of larger or more reactive molecules into smaller or less reactive ones. For example, nitrogen dioxide and volatile organic compounds in exhaust fumes can be degraded into more benign compounds such as nitrates, small carbon containing compounds and water; all of which are more easily absorbed into the environment.

Examples of this can be already be found in commercial products: Pilkington Glass have coated the surfaces of their glass with titanium dioxide nanoparticles to break down dirt. The process is started by sunlight and is completed by rainwater, which washes the residue off the surface of the glass the process and catalyses the break down of dirt. The rain is then able to easily wash it off.

New composite concretes incorporating titanium dioxide nanoparticles have been manufactured by Italcementi and used in a number of building projects, such as those pictured above. These materials break down atmospheric pollutants and have the added benefit of staying clean, seen in the spotless white buildings pictured above.

In a similar way, the new Ecopaint from Millenium Chemicals incorporates titanium dioxide nanoparticles, and can be used to coat building exteriors, fences etc to break down atmospheric pollutants

Self-cleaning glass has been described as an impossible dream. Yet, following an intensive research and development program by Pilkington—inventors of the universally used float glass process, and the world's leading glass manufacturer—new Pilkington Activ™ does just that.

Its unique dual-action uses the forces of nature to help keep the glass free from organic dirt, giving you not only the practical benefit of less cleaning, but also clearer, better-looking windows. Pilkington Activ™ is an ordinary glass with a special surface on the outside that has the unique dual-action. Once exposed to daylight, the surface chemically reacts in two ways. First, it breaks down any organic dirt deposits and second, rain water 'sheets' down the glass to wash the loosened dirt away.

From certain angles it has a slightly greater mirror effect than ordinary glass, with a faint blue tint. Otherwise, the glass is just like any other. The PhotoActiv(tm) surface is an integral part of the glass itself, so it can only be affected if the glass itself is damaged; for example, by pointed objects, abrasive cleaners or steel wool. Our tests have also shown it will not flake off or discolor. Tests have shown that the continuous-cleaning surface will last as long as the glass itself. Yes. The surface contains harmless chemical substances already found in the home, in such things as toothpaste and paint. In fact, with only small amounts of cleaning agents needed, Pilkington Activ™ self-cleaning glass is kinder to the environment than ordinary glass.

The photocatalytic action of pilkington Active™ Self-Cleaning Glass (upper) gradually breaks down and loosens dirt, while the hydrophilic surface (lower) causes rain to sheet on the glass, leaving a clean exterior surface with minimal spotting or streaking.

The secret of Pilkington Activ™ lies in its special built-in, continuous-cleaning surface, which works in two stages:

Using a 'photocatalytic' process, the PhotoActiv™ surface reacts with ultra-violet rays from natural daylight to break down and disintegrate organic dirt.

After the window has been installed, the surface of the glass takes about five days to activate itself. For the best results, after installation and initial cleaning, by the installer, the windows should be left until the coating activation period is completed before any cleaning is undertaken.

Pilkington Activ™ glass will eventually break down even heavy deposits. However, if the surface is so dirty that UV light cannot reach the glass, the self-cleaning action will not take effect. In such cases, clean the glass with warm soapy water and a soft cloth, and in a few days the process will have re-activated.

Normally, rainfall or hosing with water will be enough to keep your windows clean. In dry spells though, your windows can be washed off with a soft cloth and warm soapy water.

Activ Glass needs only a small amount of UV radiation to activate the coating—so it works on overcast days and the water sheeting effect will last through the night.Washing dirt away The second part of the process happens when rain or water hits the glass. Because Pilkington Activ™ is 'hydrophilic', instead of forming droplets the water spreads evenly over the surface, and as it runs off takes the dirt with it. Compared with conventional glass, the water also dries off very quickly and does not leave unsightly 'drying spots'. The PhotoActiv™ surface works continuously, with dirt being washed away whenever it rains. After the window has been installed, the surface of the glass takes about five days to activate itself. For the best results, after installation and initial cleaning, by the installer, the windows should be left until the coating activation period is completed before any cleaning is undertaken. Pilkington Activ™ glass will eventually break down even heavy deposits. However, if the surface is so dirty that UV light cannot reach the glass, the self-cleaning action will not take effect. In such cases, clean the glass with warm soapy water and a soft cloth, and in a few days the process will have re-activated.

Normally, rainfall or hosing with water will be enough to keep your windows clean. In dry spells though, your windows can be washed off with a soft cloth and warm soapy water.

Activ Glass needs only a small amount of UV radiation to activate the coating—so it works on overcast days and the water sheeting effect will last through the night.

Coatings for Glazing

'Smart' windows will soon be on the market as a result of advances in nanotechnology. These windows are designed to react and adapt to their environment. One such example is a glass coating which will block any excessive heat; particularly beneficial in the warm Summer months when air-conditioning costs can be expensive.

Of course, tinted windows which block the sun's rays have been available for years. However, they block the light as well as the heat, which is far from ideal. Now, thin nano-coatings of vanadium dioxide mixed with just 1.9% tungsten metal can be applied to windows and act as heat-reflectors, whilst still letting all visible light through. The nano thickness and mixture of the coating can be altered to determine the exact temperature at which heat is reflected at. This means offices and homes can remain cool without the excessive use of expensive air-conditioning, dramatically reducing costs—both financial and environmental.

Application Areas for Informatics

Material/ Technique	Applications	Time-scale (to Market Launch)
Pre 2015		
Quantum well structures	Telecommunications/optics industry. Potentially very important applications in laser development for the data communications sector.	Quantum well lasers already utilised in CD players. Not yet optimised for the communications market (i.e. fibre optics): 4–5 years.
Quantum dot structures	The aim is to use fibre optic communications in building and computers. The problems are cost and high temperature operating conditions. Quantum well/dot structures can potentially solve this problem	Quantum dots still in research stage. 7–8 years.
Photonic crystal technologies	Optical communication sector, i.e. fibre optics. Photonic integrated circuits can be nearly a million times denser than electronic ones. Their tighter confinement and novel dispersion properties also open up opportunities for very low power devices	Still in basic R&D, but very strong commercial interest emerging.

(*Contd.*)

(*Contd.*)

Material/ Technique	Applications	Time-scale (to Market Launch)
Carbon nanotubes in nano-electronics. These holdpromise as basic components for nano-electronics they can act as conductors, semi-conductors and insulators	Memory and storage; commercial prototype nanotube-based (non-volatile); RAM; display technologies; E-paper.	Commercial prototype nanotube-based RAM predicted in 1-2 years. Consumer flat screen Limited commerciali-sation of E-paper in 1-2 years.
Spintronics—the utilisation of electron spin for significantly enhanced or fundamentally new device functionality.	Ultra-high capacity disk drives and computer memories.	A read head has been demonstrated that can deal with storage densities of a terabit per square inch. In 2001 Fuji announced a new magnetic coating promising 3 gigabyte floppy disk.
Polymers	Display technologies—this sector is driven by the electronics consumer market.	Some commercialisation, e.g. Cambridge Display Technologies has been formed specifically to exploit this technology.

Post 2015

Material/ Technique	Applications	Time-scale (to Market Launch)
Molecular nano-electronics (including DNA computing)	Circuits based on single molecule and single electron transistors will appear, initially in special applications.	Single atom transistor demonstrated recently. Still immature, but huge potential
Quantum information processing (QIP)	Several researchers have devised algorithms for problems that are very computationally intensive (and thus time-consuming) for existing digital computers, which could be made much faster using the physics of quantum computers. E.g. factoring large numbers (essential for cryptographic applications), searching large databases, pattern matching, simulation of molecular and quantum phenomena.	Still in pure research phase, although some US defence money has been made available.

6

TECHNOLOGIES

(A) BIO-NANOMATERIALS

Bio-nanmaterials is a nanotechnology that has a biological application. The application can be in the medical or environmental fields and involves nanotechnology that has an intimate relationship with biological organisms. The use of nanotechnology in biomedical applications remains at its infancy.

Nanomaterials in various biomedical applications include as implants, tissue engineering materials, drug delivery devices, etc.

Historically, synthetic materials have not served as sufficient implants. For example, the current average lifetime of an orthopedic implant is only 15 years. Similarly, for the vascular community, small diameter vascular grafts are only functional 25% of time past 5 years of use. Clearly, conventional materials have not invoked proper cellular responses to regenerate tissue that allows for these devices to be successful for long periods of times. In contrast, due to their ability to mimic the dimensions of constituent components of natural tissues, nanophase materials may be an exciting successful alternative. This is not only due to their ability to simulate dimensions of proteins that comprise tissues, but also because of their higher reactivity for interactions of proteins that control cell adhesion and, thus, the ability to regenerate tissues.

Nano-Biology

The objectives of the planned research in Nanobiology are the application of nanotools to relevant medical and biological problems the refinement of such applications and the development of new tools for medicine and biology.

Major topics concern imaging of native biomolecules, membranes and tissues with the AFM, the use of cantilever array sensors for genomics, proteomics and diagnostics, and the application of nanooptics for observing and manipulating molecular processes in living cells.

Progress in imaging will be achieved by the use of multifunctional cantilevers that enable to switch biological processes while observing the structural changes, and by fast AFM employing small cantilevers and fast scanners. A wide range of questions on the structure and function of the cytoskeleton, the nuclear envelope, the cell membrane and the respective individual building blocks will be addressed. Key questions concern the nanomechanical properties of the extracellular matrix (cartilage) and cytoskeletal fibers, nuclear transport, ligand and voltage induced channel gating, and the signal transduction by membrane embedded receptors.

Native disk membranes of visual rods in the murine retina revealing rows of rhodopsin dimers, which are separated by 3.5 nm. The topograph has been recorded by operating an AFM in buffer solution.

A fast developing area is related to the fragmentary understanding of regulatory networks within cells and between cells of an organism. Such networks dictate how a cell responses to external stimuli, which activate a signaling cascade and induce expression of certain proteins. The quantitative assessment of the nucleic acids involved and the proteins expressed is a difficult endeavor. Cantilever technology is an ideal platform for such analyses, because it is sensitive and allows label-free detection of specific nucleic acids (RNA) or proteins. This technology will be integrated as lab on a chip, to provide a fast and sensitive instrument for future medical diagnostics. Imaging and array technology will be combined to develop a tool with single molecule detection capability that will ultimately allow the proteome of a single cell to be assessed.

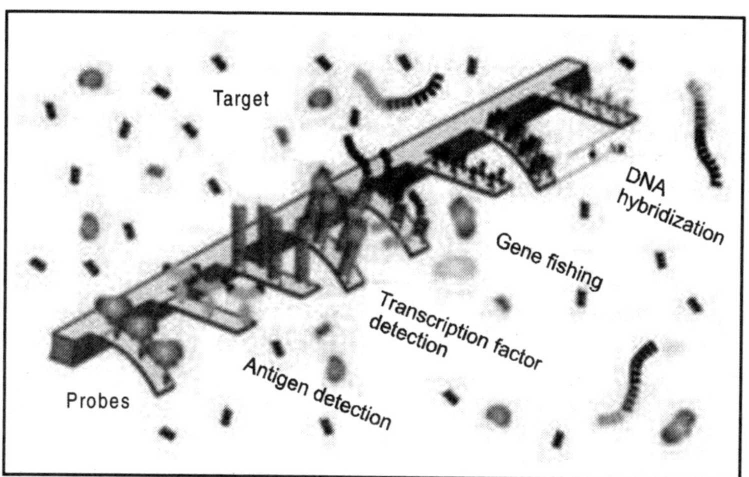

Multifunctional cantilever arrays will enable parallel in situ detection of a multitude of genomic and proteomic markers.

Nanooptics will provide tools to observe cellular events at the single molecule level and to manipulate single biomolecules for assessing their nano-mechanical properties. Such tools will be of importance in medical diagnostic, as well as in cell biology and biophysics.

Fifty years from now, what causes of death will be preventable? That depends largely on the technology we will have available, so let's start by projecting some technology trends. Gene sequencing and identification will be as easy as a blood sugar test. Medical devices such as artificial hearts and insulin pumps will be implantable and well-integrated with the body's natural demands. Surgical instruments will be more delicate and less destructive; what today is "major surgery" will be done with an office visit.

Nanotech can solve most or all of the medical problems that might keep us from being in good health, thus allowing us to remain in a state of good health for many decades or even centuries.

Intervention

Medical intervention generally consists of either surgery or drugs. To reach an area inside the body, the body must be cut somewhere. Drugs are usually delivered to the entire body at once. Most medical interventions today are designed to fix a specific problem, and are applied after the problem has already developed.

State of the art surgical technique uses instruments inserted through small tubes placed in small incisions. These instruments are necessarily simple; for example, a gripper or a blade. Although surgical robots are coming into use for certain delicate operations, the robots are considerably bigger than the area they operate on. We don't yet have robots that could fit through the tubes and do complicated operations on-site. Nanotech can eliminate this problem. The smallest acupuncture needle is 120 microns, or about as wide as twelve cells. 120 microns is 2,400 times as wide as a flagellar or electrostatic motor. A remarkably complex surgical robot could thus be inserted through a hole so small it doesn't even bleed.

Nanobots will probably be able to stitch tissue together at a cellular/molecular level, greatly accelerating the wound-healing process. This means that if large incisions are required. for example to replace whole organs, they can be repaired as part of the surgery. Accidental trauma will also be relatively easy to fix.

The normal way to deliver a chemical today is to dump it into either the bloodstream or the stomach, and let it spread all through the body. For some chemicals, such as insulin, this is appropriate. But for others, such as chemotherapy drugs and some antibiotics, it is best to keep them as local as possible. Nanosurgical techniques can put drug delivery devices right where they are needed. The devices can be numerous and tiny, so that they can be inserted into any organ. In most cases, the devices could manufacture the required chemicals on the spot, using elements and energy from the surrounding tissue, thus eliminating the need for holding tanks and external supply.

Replacement

If an organ fails, we must either replace it or do without. Usually the replacement organ comes from someone else, which means that the body will reject it unless drugs are taken to cripple the immune system. Today several organs, including the larynx and the bladder, have been grown on special scaffolding. With nanotech to build far more complex and precise scaffolding, we will be able to create most organs this way from the patient's own cells, thus allowing rejection-free transplantation.

Artificial organs will become far more feasible. Today, artificial hearts have been used in a few cases, and the use of external artificial kidneys (dialysis) is common. These devices don't work very well, though they are certainly better than nothing. However, a nanotech-built device could use the body's own energy supply—glucose and oxygen—for power, and could be far more sensitive and responsive to the body's condition

Repair

Today, if a tissue is torn or cut, we must simply wait for the body to repair it. The most we can do to help is to hold the torn edges together with stitches or surgical glue. (nanobots,a robot with parts of atomic scale and precision). A typical nanobot, smaller than a cell, may contain billions of atoms. But nanobots need not be small. Sometimes it's useful to fit a lot of functionality into a single device. Some nanobots may have structural application, such as reinforcing bone or replacing muscle—such devices could be quite large. The important point about a nanobot is that it can make efficient use of space, with functionality as densely packed as a bacterium) should be able to re-form the molecular bonds that hold cells together, and thus repair wounds almost immediately.

Another form of injury is oxygen deprivation. Due to a blood clot or broken blood vessel, a tissue may be starved of oxygen. Normally this causes cells to kill themselves within minutes.

However, drugs have already been found that tell the cells not to give up so soon; in early trials, they seem to cause significant improvement in stroke victims. A population of nanobots scattered throughout the tissues could provide more timely and targeted release of such drugs, and could also store a few minutes worth of oxygen in pressurized tanks to keep the tissue alive until the wound repair machines can fix the problem.

Heterostasis

The body maintains its condition by a mechanism called homeostasis. There are hundreds of signals controlling hundreds of mechanisms, so that if part of the body starts to get out of sync with the others it is forced back in line. For the most part, these signals are sent by either chemical or neural signals. There is no overall control, and some of the signals cause undesired side effects.

Heterostasis is the idea that different parts of the body can be maintained deliberately out of sync with each other. For example, it appears that some immune diseases such as asthma may be caused by a lack of parasites. At least one doctor has deliberately infected himself with tapeworm in an effort to improve his immune function. Rather than go to such lengths, it may be possible to modify local chemical concentrations and/or the body's sensors for those chemicals, so that different systems have a slightly different picture of what's going on. It will take a lot of research to find what combinations of state are best, but it seems clear that our bodies are naturally optimized for a lifestyle different from the one we have chosen, and heterostasis may be a way to improve health.

Heterostasis may also be useful when modifying individual organs. Rather than trying to design a new organ to function precisely like the one it replaces, it may be easier to tweak the body's other systems so that they react correctly to the change. This also raises the possibility of maintaining different organs at different physiological ages for peak performance—a 90-year-old person might be healthiest with a ten-year-old liver but a 25-year-old heart.

The most extreme type of heterostasis would involve the separation of the body's components into independent subsystems, temporarily preventing all signaling between them. This would be an aggressive but straightforward treatment for massive damage or other dysfunction, in which part of the body was damaged enough to make the rest ill. Today, we can keep many organs alive for hours or even days outside the body, and we can keep the body alive for hours on a heart-lung machine. Our technology is incredibly crude in comparison with a nanotech-built interface that could simulate a healthy body in great detail. Each organ or system could thus be stabilized and repaired (or replaced) individually, without any harmful or unexpected messages from the other organs. Once everything was working well, the state of each organ would be synchronized, connections would be restored, and the body would be whole again. In December of 2001, a surgical team removed a person's liver, carried it to a nuclear reactor for anti-cancer radiation therapy, and then put it back in; a year later, the patient is alive and well.

Diseases and Cures

Medical science has scored some impressive successes. Diseases caused by bacteria have been greatly reduced by antibiotics. Vitamin and mineral deficiency diseases are almost unknown, at least in developed nations. However, we still have many diseases that limit our lifespan, and that medicine can only postpone, not cure. Life cannot be extended indefinitely without curing each disease that threatens to shorten it. Let us see some problems and how nanotech can be used to cure them.

Telomere Loss

Most cells have a length of DNA called the "telomere" that gets shorter each time they divide. After a certain number of divisions, the telomere is gone, and they die. (This is probably an anti-cancer mechanism.) If life is to be extended, cells will need to have their telomeres replaced so that they can keep working. We know that cancer cells have managed to avoid the telomere trap, and we already know of an enzyme that performs this function. It should be simple to induce a cell to lengthen its telomeres, using a machine built on the same scale as the cell that can sense its state and dispense the right chemicals at the right time.

Chemical Accumulation

One cause of cell death is accumulation of harmful chemicals. The most famous type of chemical is the prion, a malformed protein that cannot be removed by the body and that causes normal protein to turn into prions. Prions are responsible for Mad Cow disease and similar human diseases. It is unclear how many other problems may be caused by the accumulation of other non-digestible chemicals. What is clear is that a diamond nanobot could make short work of breaking up a prion, or any other chemical that the body couldn't deal with on its own. Nanobots could go from cell to cell like a housecleaning service, absorbing and breaking down a variety of undesired chemicals.

DNA Damage

Our genetic material is under constant attack from radiation and chemicals. Damage accumulates and causes cells to malfunction. This can be corrected in several ways. First, cells other than neurons that are malfunctioning can usually be killed; the body will replace them with no ill effects. In fact, cells contain several mechanisms for killing themselves if they detect that they are not working right. (Stem cells and other techniques can help if the body is slow to replace the missing cells.) Second, it should be possible to minimize damage by vacuuming up the chemicals that cause mutation, and by manipulating the cell's state to increase the amount of energy it spends on self-repair. Third, a nanomachine may scan each cell's DNA to search for and repair damage, or perhaps simply replace chromosomes periodically with new error-free copies.

Cancer

At a cellular level, cancers are usually quite different from normal tissue. Many cancer cells actually change the chemicals on their surface, so are easy to identify. Most of the cells grow faster or change shape. And every cancer involves a genetic change that causes a difference in the chemicals inside the cell.

The immune system already takes advantage of surface markers to destroy cancer cells; however, this is not enough to keep us cancer-free. Nanobots will have several advantages. First, they can physically enter cells and scan the chemicals inside. Second, they can have onboard computers that allow them to do calculations not available to immune cells. Third, nanobots can be programmed and deployed after a cancer is diagnosed, whereas the immune system is always guessing about whether a cancer exists. Nanobots can scan each of the body's cells for cancerous tendencies, and subject any suspicious cells to careful analysis; if a cancer is detected, they can wipe it out quickly, using more focused and vigorous tactics than the immune system is designed for.

Brain Damage

The brain is unique among the body's organs: it stores our memories and personality, so that it cannot simply be replaced if it starts to wear out. This poses a special problem for life extension: the information stored in the brain must be preserved over extended periods of time, safe from disease and accident.

Obviously it is good to prevent the premature death of neurons. Poisons such as alcohol, accidents such as stroke, and diseases such as Alzheimer's can all cause neurons to die. In each of these cases, neuron death can be greatly slowed if not prevented entirely by controlling the chemistry inside the cell. Injurious chemicals can be vacuumed up and converted into harmless ones. Damaged neurons, like other cells, sometimes go into suicide mode (called "apoptosis"); as mentioned above, this can be chemically prevented, and the neuron can be stabilized until the problem is fixed and the damage is repaired.

It is now known that brain cells do regenerate: the brain is adding new ones all the time. This implies that some neural death is normal. How do the new cells know how to behave? It seems that a new neuron can take its cues from the existing ones; this means that a person's mind may be intact even after the death and replacement of a large percentage of their neurons.

Finally, it may be possible to measure neural connections and/ or activity in enough detail to simulate the firing pattern. This may make it possible to create an artificial neuron or even an artificial neural net that can be used to replace missing neurons and retain old memories. But even if this proves to be impossible, the worst-case scenario is one in which people can't remember much farther than a century back. We accept more memory loss than this as a natural consequence of aging.

Hormone Deficiency

Aging is associated with changes in the levels of many hormones; perhaps the best known example is menopause, which is caused by a reduction in estrogen. It is likely that treating glands against aging at the cellular level would restore age-appropriate hormone production. However, if this is not enough to bring the body to a younger state, artificial glands could be built that would maintain the desired hormone levels. In fact, different hormone levels could be supplied to different organs—something that the body cannot do for itself. This would be an example of heterostasis.

Infection

Bacteria, viruses, and parasites are continuing problems. Antibiotics work well against most bacteria; however, antibiotic-resistant strains are developing. Since viruses aren't active until they take over a cell, they are immune to antibiotics, and medicine cannot yet do much against them. There are many kinds of parasites that may need individual medical techniques.

Our immune system is quite effective at dealing with most infections. However, it needs to learn by experience—it is generally most effective at fighting organisms that it can recognize on a molecular level. Diseases can be very clever in evading it. Some diseases, such as Ebola, progress too rapidly for the immune system to respond. Syphilis survives by being stealthy and surrounding itself with the body's own chemicals to camouflage itself. Herpes splices itself into the genes of the body's cells, so the immune system can't detect it and wipe it out. HIV directly attacks the immune system.

Nanobots have several advantages over the immune system. They will not be susceptible to attack by natural pathogens. They will have computational resources unavailable to immune cells. They can be programmed to find and fight diseases they have never encountered—when a new disease shows up, as soon as it is analyzed. Likewise, the system can be activated based on external knowledge of the likelihood of a disease; the nanobots won't have to waste energy looking for malaria in winter. Nanotech will give us more options for cleaning up after a disease, since corrupted genes will be repairable without killing the affected cell.

Some diseases, such as cholera and tetanus, live in the environment; without scrubbing the whole earth, we can't get rid of them entirely, so we will need to maintain an immune system against them. But many diseases can't survive without humans to infect. With great effort, western countries managed to eradicate smallpox using 1970's technology. Cheap manufacturing would allow the creation of billions of doses of highly effective treatments that would be easy to distribute and administer; the main obstacles to wiping out many diseases worldwide would be political, not economic or technological.

Accidents

Accidents, especially motor vehicle accidents, are a leading cause of death at all ages. Although an accident is not itself a disease, it kills by producing damage to the body, and that damage can be treated or prevented like any other disease. Most accidents involve mechanical injury (trauma); most of the rest involve chemical injury, either poisoning or oxygen starvation. A permanent nanobot installation can make many accidents survivable that would be fatal today.

Nanobots embedded in tissue can strengthen it against tearing, or repair it if it does tear. It is common for a blow to the head to rattle the brain against the skull; a specially shaped nano-built device could cushion the brain, preventing this damage. Other devices could vacuum up common poisons before they could cause damage, or barricade poisoned areas to keep the poison from spreading through the body. Respirocytes could allow the body to function normally for several minutes without breathing or circulation, giving more opportunity to restore normal functioning. In cases of extreme injury, heterostasis could be used to stabilize the body until help can arrive. As long as the brain is not physically damaged, it can be functionally separated from the body and forced into a low-power state. With today's medicine, paramedics refer to the "golden hour": if an accident victim can be brought to a hospital in less than an hour, chance of survival is greatly increased. People have recovered after drowning in cold water for over an hour; artificial mimicry of this state, combined with the ability to aggressively repair the body, might extend the "golden hour" significantly.

Blood-Related Diseases

Many diseases, from heart attacks and strokes to sepsis and metastasizing cancer, involve the blood in some way. A nanomedical device, a "Vasculoid" would replace the blood volume and take over its functions by lining the entire vascular system with a multi-segmented robot. In addition to preventing many diseases, and limiting the scope of others (such as poisoning), such a system would provide detailed control of the body's chemical environment around each individual capillary, allowing heterostasis to be used extensively.

The Vasculoid is extremely complicated and would require much research to build and use successfully. This particular device may never be used, but it can provide a hint of the possibilities inherent in advanced nanomedicine

Current Implant Failures

First, we will define current problems with various implants and learn why current materials fail in the device systems.

Orthopedic/Dental

- Current Materials Used
- Current Modes of Failure: Poor bonding to surrounding bone; Generation of wear debris; Stress and strain imbalances at the tissue implant interface
- Current Statistics on Orthopedic Implant Failure

Cartilage

- Current Modes of Failure: Poor regeneration of cartilage tissue; Generation of wear debris; Stress and strain imbalances at the tissue implant interface.

Vascular

- Current Modes of Failure: Poor bonding to surrounding vascular; Endothelium removal due to shear stress; Generation of stenosis; Stress and strain imbalances at the tissue implant interface.

Bladder

- Current Modes of Failure: Poor bonding to surrounding tissue; Stress and strain imbalances at the tissue implant interface

Central and Peripheral Nervous System

- Current Modes of Failure: Poor bonding to surrounding tissue; Build-up of glial scar tissue; Poor maintenance of electrical properties; Stress and strain imbalances at the tissue implant interface.

Biological Response of Implanted Materials

Next, we will learn how cells interact with implanted materials to generate a biological response. We will discuss both desirable and undesirable reactions of the body with implanted materials.

Protein Interactions with Implanted Materials

- Properties of Proteins
- Adsorption
- Conformation/Bioactivity

Cellular Recognition of Proteins Adsorbed on Materials Surfaces

- Adhesion
- Migration
- Differentiation

Cellular Extracellular Matrix Deposition Leading to Tissue Regeneration

- Bone
- Dental Tissue

- Cartilage
- Vascular
- Bladder
- Central and Peripheral Nervous System

Foreign-body Response

- Inflammatory Response
- Wear Debris Response

Advantages of Nanomaterial Use as Implants

How can nanophase materials be used as better implants? What are their promising properties and what experimental evidence exists?

Mechanical Properties

- Ceramics: Increased grain boundary sliding
- Metals: Dislocation source
- Polymer Composites

Electrical

Surface

- Biologically-inspired Surface Roughness
- Increased Surface Reactivity

Multi-plex Biomolecular Detection Using Bio-nanotransduction

Bio-nanotransduction is a biological detection scheme that is based on biological recognition (Bio) and the production of biological nano-signals which allows the change of recognition event into universal signal molecules (nanotransduction). A single sample can be converted into a pool of nano-signals specific for each target. Detection and differentiation of the nano-signals are correlated to the presence of targets in the original sample. It is possible to link various biological recognition events to platforms (such as biosensors) capable of detecting nano-signals. This allows the application of new target recognition while using a universal platform previously optimized for nano-signal detection.

Detection of Biological Species by Surface Enhanced Raman Scattering

The rationale behind this work is to explore the application of surface-enhanced Raman scattering (SERS) to detect and identify biological species, especially some common pathogens that are involved in food poisoning and water contamination. SERS is a powerful tool to probe bacterial cell wall surfaces and provides rich chemical information of the different chemical moieties present in the cell wall. SERS also provides a label-free optical detection technique, which is non-invasive and less time consuming when compared to the conventional methods where we need to grow the bacterial cells over a period of 18-24 hours prior to analysis. In this work we explored the identification and detection of bacteria by SERS using nanometallic particles. The detection limit for our system is at least two orders of magnitude lower than the FDA recommended value. Another objective of our research is to develop an online SERS based detector system for water and airborne biological species. In this respect, we were able to obtain reproducible bacterial fingerprints after aerosolizing them from an aqueous suspension.

Integrated Techniques for Transmembrane Protein Sensing

Method for the microfabrication of apertures in a silicon substrate using well-known cleanroom technologies and a novel approach using phase sensitive detection to measure single channels of transmembrane proteins. These proteins are of considerable interest because they are fundamental to sending signals throughout the human body. Microfabrication of parallel apertures of such devices could be used for high throughput screening methodologies in both pharmaceutical and integrated biological sensor applications. Using engineered proteins and the presented integrated device, a sensor capable of detecting single molecular analytes can be envisioned. Our results demonstrate a working fabricated silicon device that is capable of such measurements. We also demonstrate a lock-in amplifier technique used to record single channels of transmembrane proteins inserted into bilayer membranes formed on the silicon device.

Biomarker Discovery for Exposure of Macrophage Cells to Nanoparticles

Harmful effects of particles to macrophage cells depend on their surface area and intrinsic toxicity of chemicals absorbed on the surface. As particles become smaller, it is more likely they will harm the lung. For example, nanoparticles may gain access to sites, which larger particles cannot reach. Increasing commercialization and use of nanomaterials will result in increased exposure to nanoparticles through inhalation, ingestion, skin uptake, and injection of engineered nanomaterials. The complexity of issues relating nanoparticle composition, size, deposition in the lung, and cellular responses necessitates monitoring the health impacts and developing methods that can capture the multifaceted nature of this problem. We are developing capabilities to monitor health effects of nanoparticles in a variety of cell types using live cell infrared (IR) [e.g., attenuated total reflectance (ATR)] spectroscopy and fluorescence imaging techniques. In situ IR spectroscopy has desirable features since this technique does not require knowledge of the biological endpoint to monitor a priori and is not dependent on contrast agents for detection of the biological response. ATR measurements require the long term culture of cells on an IR-transparent substrate in a spatially restrictive manner. Specifically, ATR measurements capture information localized within about 1nm off the crystal surface. We have engineered an FT-IR system for live cell monitoring using ATR spectroscopy. Researcher examined dynamic changes in ATR spectra associated with treatment of RAW 264.7 macrophage-like cells with bacterial endotoxin (lipopolysaccharide, LPS) and carbon nanotubes. Specific chemical bond changes at a variety of wave numbers were observed. Based on our initial investigations, researcher are optimistic that combinatorial approach can contribute to understand the health effects of nanoparticles.

The Effect of Gold Nanoparticles on Dendritic Cells

Recent biological applications have been focusing on using gold nanoparticles for e.g. drug and gene delivery. There is however little information available concerning what influence such particles have on the immune system, e.g. on dendritic cells (DCs).

DCs are the main form of professional antigen-presenting cells having a unique ability to induce both primary and secondary immune responses by expressing cytkines, MHC and co-stimulatory molecules such as CD80, CD83 and CD86.

Researcher addressed the question whether gold nanoparticles of 6 nm affect DCs, looking at morphology, viability, expression of cytokines and of co-stimulatory and antigen presenting molecules. This was assessed by using human DCs together with gold nanoparticles, and various techniques including microscopy, flow cytometry and ELISpot.

Experiments revealed that both morphology and viability are almost not affected by the gold nanoparticles. The expression of CD40, CD80, CD83, CD86 and MHC class I and II was not or only to a minor degree upregulated, which also confirms the biocompatibility of gold. This is further supported by low or no expression of the cytokines IL-10, IL-12 and INF-alpha. In conclusion, gold nanoparticles of 6 nm in diameter have the potential to be used as inert carriers in biomedical applications.

The Use of Micro and Nanoparticle Zinc Oxide in Personal Care Products Based on Skin Penetration

Nanoparticles of ZnO and TiO_2 are increasingly being used as pigments and UV absorbers in personal care products (e.g. sunscreens), coatings and paints, predominantly because their UV absorbance efficiency and transparency to visible light increases with decreasing particle size. Receiving particular attention is the use of nanoparticles in personal care products with the main concern being their phototoxicology. Scant attention has been placed on the effect of particle size on the fate of these materials after application including their penetration into and possibly through the skin.

To assess and compare the in vivo penetration behaviour of nano- and micro- sized ZnO particulates into human stratum corneum (SC) after topical sunscreen application and to determine the penetration of nano-sized ZnO into the SC and hence its distribution on or in the uppermost layers of the SC, a method of sequential removal of layers of stratum corneum by adhesive tapes known as tape stripping was adopted.

The study revealed that a large proportion of the ZnO applied remained on the surface or at least in the uppermost layers of the SC and the mean percentage of recovery after 18 tape strips varied between approximately 70 and 100%. The differences in the recovery between nano and micro ZnO at both $t = 0$ and $t = 2$ hrs were insignificant ($p > 0.05$) suggesting the reduced particle size of ZnO alone did not enhance its ability to penetrate further into the SC.

Although the study suggested that there was little apparent difference in the penetration behaviour of these particles into the stratum corneum, it raised several potentially significant issues relating to the efficacy of using nanosized additives in personal care applications. These will be discussed.

Development of Peptide Conjugated Superparamagnetic Iron Oxide (SPIO) Nanoparticles for Targeted MR Imaging and Therapy of Pancreatic Cancer

Researcher have engineered superparamagnetic iron oxide (IO) nanoparticles targeted to urokinase plasminogen receptors (uPAR), which is highly expressed in tumor cells, intratumoral fibroblasts and tumor endothelial cells and can also be internalized by cells. IO nanoparticles with highly uniformed core sizes were coated with amphiphilic polymers and conjugated to peptides containing a uPAR binding region of uPA (ATF-IO). Researcher demonstrated that ATF-IO nanoparticles specifically bound to human pancreatic cancer cells evidenced by significantly shortened T2 in ATF-IO-treated cells but not in unconjugated IO or GFP-IO. MRI scan of nude mice bearing orthotopic pancreatic tumors before IO particle injection showed bright signals in the tumor areas. However, after systemic delivery of ATF-IO particles into the mouse, we observed a significant signal drop in those areas, suggesting a T2 shortening effect of selective accumulation of ATF-IO in pancreatic cancers. However, MRI images from a tumor-bearing mouse injected with unconjugated IO particles did not show a significant T2 effect in the tumor area. Our results demonstrated that ATF-IO nanoparticles have great potential for the development of multi-functional nanoparticles for in vivo imaging and therapy of pancreatic cancer.

Detection of Live Breast Cancer Cells Using Carbon Nanotube Devices

Development of new technologies for reliable early detection of cancer from biological fluids via minimally invasive methods is still a high priority. Cancer cells often overexpress characteristic surface receptors, which provide an opportunity for early diagnosis of disease. Monoclonal antibodies specific to cell surface antigens overexpressed on cancer cells can be absorbed to single wall carbon nanotube (SWNT) devices, resulting in a drop in conductance. Application of human BT474 and MCF7 breast cancer cells increased the conductance of the insulin-like growth factor 1 receptor (IGF1R) specific mAB-SWNT devices. The degree of increase in conductance of the devices due to live cancer cell application was proportional to the number of overexpressed surface receptors in cancer cells

The Targeted Drug Delivery of Doxorubicin

Carigent has developed a novel technology for attaching a high density of biological molecules to the surface of biodegradable, drug-loaded PLGA particles. These particles are characterized by a long-lasting presentation of the high density surface ligands. The combination of these factors position the Carigent technology as a premier reformulation apparatus for a large variety of drugs and target ligands. Because of this flexibility and ease of preparation this technology promises unprecedented effectiveness in a range of clinical therapeutic applications for cancer, cardiovascular, autoimmune and alloimmune disease.

An exciting aspect of Carigent's technology is its versatility. Any drug, small or large molecule can be encapsulated inside of the particles while any desired ligand molecule can be attached to the surface. Carigent is currently focused on several products using this technology including a targeted version of doxorubicin for anti-cancer (non-hodgkins lymphoma) treatment. The doxorubicin re-formulation shows one example of the power of this technique; pre-clinical experimentation has shown that Carigent targeted doxorubicin is 100 times more powerful and with lower toxicitythan free doxorubicin.

This versatility will offer ample opportunity for corporate licensing and partnering with pharmaceutical and biotechnology companies for further generic drug reformulation.

In Vivo Tracking of Liposomes Using CT and MR Imaging

Currently, there are limited methods available for longitudinal and non-invasive in vivo assessment of the transport kinetics of carrier-based therapeutics, such as those relying on liposomes. A viable strategy to non-invasively track the carrier in vivo is to employ the same carrier loaded with contrast-enhancing agents such as iodine and gadolinium. If these agents are retained within the carrier, they may act as surrogates for therapeutic molecules and allow for non-invasive tracking of the carrier in vivo using computed tomography (CT) and magnetic resonance (MR) imaging devices. If a correlation can be found between the change in signal intensities measured with CT and/or MR and the actual agent concentration detected in vivo, then this method has the potential to be successfully adopted for non-invasive quantification of pharmacokinetics and biodistribution studies. In this way, not only can the same animal be sampled over multiple time points, avoiding animal-to-animal variations, but the total number of animals required for each study can also be drastically reduced. Specifically, this study evaluated the in vivo pharmacokinetics, in both mice and rabbits, of a liposome-based CT and MR contrast agent previously developed by our research group. Correlations were established between the iodine and gadolinium concentrations found in the rabbit plasma by HPLC and ICP assays and the signal enhancement obtained in the rabbit aorta in CT and MR using circular regions of interest over all time points. A linear relationship was found between the iodine concentration and Hounsfield Unit increase in CT, while an exponential relationship was observed between the gadolinium concentration and signal intensity increase in MR. This pilot study demonstrated the potential of employing CT and MR imaging for longitudinal in vivo mapping of contrast agent containing liposomes.

Nanoporous Silicon Sensor for Biological Applications

The optical and electrical parameters of nanoporous silicon fabricated with heavily doped p-type silicon are very sensitive to the gases and organic solvents. For sensing the gases and biomaterials the two types of sensors were developed. The strong correlation between structural properties (pore morphology and nanoporous silicon layer thickness) and the sensitivity of sensors was obtained. The response of bio-molecules with different dipole moments was considered. It was shown, that conductance and capacitance is increasing, when the molecular dipole moment is increased.

Nano-approach for Sensitive Monitoring of Bacteria

The sensitive method for pathogen identification is very important in medicine, food analysis, and environmental monitoring. The unique ability of AFM to image bacterial world at subnanometer resolution and in aqueous solution was opened new fields to investigate bacteria themselves and their surfaces under physiological conditions without any type of labeling. In the present work, we attempt to identify pathogen bacteria Legionella pneumophilla and bacterial spores of Bacillus thurengiensis with using AFM. For specific bacterial visualization the new approach based on immobilization of specific antibodies indirectly to the solid surface via antibody-binding staphylococcal protein A was developed. Bacterial cells and spores have been chosen to discriminate objects on AFM pictures by forms and sizes in the presence contaminates. The layer of antibodies against bacteria was exposed with bacteria and to evaluate the specificity with spores. The layer of antibodies against spores was exposed with spores and bacteria too. The AFM pictures were analyzed on the presence bacterial cells or spores.

Neuroimaging Distribution of Water Molecules in Brain

To elucidate and build the map of the water molecules distribution in a brain tissue in hippocampus/lateral vetricle area by means of the injected chromophore bis-N-carbazolyl-distyrylbenzene (BND) and FLIM technique. The BND is highly eficient two-photon absorbing chromophore. The fluorescence lifetime of BND diminishes when the amount of water in the surrounding increases.

Researcher observed different fluorescence lifetime of the injected in vivo chromophore depending on water content in different areas: hippocampus (extracellular fluid), lateral ventricle (choroid plexus) (cerebrospinal fluid); and blood vessels whithin the areas. Moreover, the fluorescence lifetime distribution had been drastically changed in case of water deprivation compared to the normal drinking conditions. Statistical analysis of the investigated samples showed that water content decreased at a border of the hippocampus/lateral ventricle area in a deprived case and increased in blood vessels.

The present research involves a detailed investigation of the mechanism of chromophores—water interaction; visualization of water molecules redistribution in brain caused by dehydration; and as results—preliminary investigation of mechanisms of brain self-protection from water deprivation.

In Vitro Human Model for Tumor Target Screening i n Cancer Research

Monolayer and suspension cultures are still a highly artificial cellular environment for target screening and drug development. They have limited predictive value for target screening especially for nanomaterials. Target screening requires test systems that mimic the human cancer with increasing accuracy in order to optimize the selection of potential effectors.

Tissue engineering is not only the generation of tissues it should also indentify biological principles of cellular behaviour.

Researcher developed a new 3D vascularised test system. This system is based on decellularized porcine small bowl segments and preserved tubular structures of the capillary network within the collagen matrix which is functional associated with one small vein and artery (vascularised matrix). This vascularised matrix enables the generation of a functional artificial vascular network. Therefore the vascular network is populated with endothelial cells. Stain for vitality and the specific endothelial markers such as CD31, VE-Cadherin and Flk-1 matched functional finding by Insulin dependent FDG uptake predominantly in the region of the former vascular structures. This bioartificial vascularised matrix can be additional populated with tumour cells to create ex vivo vascularised tumour-like structures.

This model offers the possibility to simulate physiological drug application and a human 3D testsystem to established nanomaterials/systems for cancer research/therapy.

Design of a Wireless Passive Microsensor for Injection into the Human Brain

It is powered and communicates by induction from an external coil, and incorporates a novel design of 'micro umbrella' to mechanically deploy micro inductors on 'wings' as the sensor exits the needle to increase the communication capacity.

Preservation of DNA and Protein Biofunctionality for Bio-MEMS and NEMS Fabrication

Microfabrication processes especially in silicon are not compatible with biomolecules. For example, silicon processing is commonly performed at high temperatures and makes use of aggressive chemicals. Under these conditions biological materials are denaturated or decomposed. Currently biological materials can only be combined with microdevices after the last microfabrication production step.

Researcher report the integration of DNA and protein into the microfabrication process using a gold passivation mask preserving their biofunctionality. This technology enables the integration of DNA and potentially other biomolecules into mass scalable microfabrication processes for biomedical devices, biochips, and biosensors. The gold film served as a biocompatible and stable passivation mask in the subsequent microfabrication processes. Biomolecules were patterned and aligned with microstructures formed by wet etching and microelectrodes created by a metal deposition/lift-off process. After microfabrication the gold passivation mask was removed under mild conditions to uncover the DNA or proteins. A prototype device was fabricated and biofunctionality of DNA was almost completely preserverd, demonstrating the full integration of DNA into silicon microfabrication for the first time.

Reflex-arc on a Chip: An in Silico Cell Culture Analogue

To overcome the shortcomings of in vitro and animal models researcher are developing a microscopic cell culture analogue (microCCA) of the spinal reflex-arc. The present work draws on advances in a wide variety of technical fields including cell culture, surface chemistry, and microfabrication. These advances have allowed Researcher to begin development of a microCCA device comprised of the basic components of the reflex-arc: a muscle fiber, a dorsal root ganglion cell, and a motoneuron. Silicon microstructures serve as the foundation of the device. Surface modification with alkyl-silane SAMs followed by patterning with deep UV photolithography to selectively control cell adhesion and growth. Furthermore, researcher have demonstrated the control of neuronal growth and myotube differentiation on the microstructures. This system will enable the controlled interrogation of properties of the reflex-arc, thereby creating an improved test bed for the development of novel drug therapies for traumatic SCI and neurodegenerative diseases

Interaction between Artificial Mucin Layer and Stimuli-Responsive Nanogel Particles for the Oral Peptide Delivery Observed in Simulated Intestinal Solutions by Using Colloid Probe AFM Method

In order to characterize and control the adhesive behaviors of nanometer scaled stimuli-responsible gel particles designed for oral peptide delivery, their interaction with artificial mucin layer in the small intestinal solutions was determined by the colloid probe atomic force microscope method. The prepared nanometer scaled gel particles with a core-shell structure were designed to exhibit behaviors responsive to temperature and pH in solutions, consequently protect the incorporated peptide drug under harsh acidic conditions in the stomach, adhere and penetrate to the mucin layer in the small intestine, and thereafter release the drugs. Spherical agglomerates of the nano-gel particles with several micron meters in diameter were prepared by the spray freeze drying method and adhered on the top of tip of commercial atomic force microscope. The interaction between the artificial mucin layer and nano-gel surface determined by the colloid probe method depended on pH of the solution. Based on the possible transition of the surface-microstructure of nano-gel particles following the pH change and the measured results from the colloid probe AFM method.

Mitotic Inhibition by Varying the Frequency of Water Clusters

Computer simulations were developed at MIT in the 1970s to relate molecular structures to chemical & physical properties. These simulations were applied to the development of revolutionary catalysts for the petroleum industry, ferromagnets for the recording industry and new superconducting materials. Subsequently, the same approach was applied to liquid water in 1998 in an effort to understand the role of water in combustion of liquid fuels. These simulations indicated that water is not simply a random assembly of molecules, but has well-defined structural units. These structures are dynamical in that they form and break apart at a high rate, but once formed, they exhibit unusual vibrational modes referred to as anharmonic modes. The basic structural unit is a one nanometer diameter icosahedron containing one water molecule at each of its twenty vertices. These findings were supported by x-ray and Raman spectroscopy and published in the Journal, Science in 1998.

The successful simulation of water structures suggested several means for changing the frequency of the anharmonic modes and thereby impacting the catalytic nature of water. This ability to alter the vibrational properties of water has had a measurable impact on pathogens in biological systems.

Drug Delivery by Surfactants

Surfactants are amphipathic substances with lyophobic and lyophilic groups making them capable of adsorbing at the interfaces between liquids, solids and gases, to form self-associated clusters, which normally lead to organized molecular assemblies, monolayers, micelles, vesicles and membranes. Their capability of reducing surface tension allows them to mix or disperse readily as emulsions in water or other liquids. Surfactants are among the most versatile products of the chemical industry having potential applications ranging from agricultural sprays to oil recovery including areas such as catalysis, coatings, dispersions, electronics, floatation of minerals and lubrication, etc. Where as biomedical applications of surfactants extend from anesthesiology to zoology covering artificial implants, gene transfection, biomembranes, biolubrication, lipoproteins, opthalmogy, pharmaceuticals and pharmacology. Moreover, biological surfactants perform a vital role in the metabolic processes of living organisms. For instance, the pathological effects of sucrose like dental cavities and also higher caloric contents have led to the use of certain surfactants e.g. the non-ionic sorbitol as sugar substitutes in some confection. Where as, phospholipid lecithin, bile salts, certain fatty acids and their derivatives are the natural emulsifiers that by reducing the surface tension of fat globules help them to be dispersed in water and hence get solubilized. In pharmaceutical processing they have become indispensable since they afford a uniquely effective and efficient mechanism of drug carriage by solubilizing the drugs of fatty origin. In this article potential applications of some selected surfactants as drug carriers have been reviewed.

Targeting Human Brain Tumor with Anti-Nucleosome Antibody 2C5-modified Liposomes in Nude Mice Model

Delivery systems such as liposomes have shown promising potential to overcome the existing brain barriers and deliver the anti-cancer drugs in therapeutic amounts to brain tumors. Further modification of liposomes with monoclonal antibodies enhances their ability to target the desired area. Previous studies have shown the potential of monoclonal antibody 2C5 (mAb 2C5), an anti-nucleosome antibody, to target nanocarriers such as micelles and liposomes to a variety of tumors. We hypothesized that mAb 2C5 can be used for the targeted delivery of doxorubicin-loaded liposomes to brain tumors in vivo. The cytotoxicity studies with 2C5-modified Doxil® (a commercial preparation of liposomal doxorubicin) demonstrated the ability of mAb 2C5 to target the liposomes and deliver anti-cancer agent doxorubicin within U-87 MG brain tumor cells to achieve cell death at a lower dose of doxorubicin. The therapeutic treatment studies on intracranial tumor model in nude mice showed that 2C5-Doxil® was significantly effective in inhibiting the tumor growth and hence prolonging the survival time of the tumor-bearing mice over control liposomes. Researcher therefore concluded that 2C5-modified long circulating immunoliposomes can potentially be considered for the delivery of anti-cancer agents, and hence for the therapeutic treatment of human brain tumors in vivo

Sensitive Peptide-Nanotube Based Pathogen Assay

A novel peptide nanotube-based multiplexed pathogen assay has been developed that has high sensitivity, detecting as few as 100 pathogens per mL in as little as 0.5mL. Multifunctional nanotubes detect viruses by selectively binding antibody at the ends of the nanotube to the virus, while the length of the sidewall is used for signaling with fluorescent dyes. Multiple nanotubes bind one virus at low pathogen levels and begin to form three dimensional networks of many nanotubes binding many pathogens as observed by TEM. Discussion will include details on this mechanism of network formation. These single and multiple virus network structures are quantified by flow cytometry. By assigning unique fluorescent dyes to each antibody used, one can create a multiplex assay limited only by the ability to resolve each dye. Proof of principle experiments with several viruses, including Herpes Simplex Virus Type 2, Adenovirus Type 3 and Type 5 and Influenza strains A and B will be presented.

Fast and Easy Detection System of Protein Glycosylation by Using Resonant Energy Transfer between Nanoparticles

Researcher construct the new glycosylation detection system using Resonance Energy Transfer (RET) between gold nanoparticles (AuNPs) and semiconductor quantum dots (QDs). RET can be induced by biomolecular interactions between concanavalin A (conA)-conjugated AuNP (conA-AuNP) as an acceptor and dextran-conjugated QD (dex-QD) as a donor, resulting the photoluminescence (PL) quenching of QDs. The RET efficiency was changed in the presence of glycoprotein that competes with the dextran on dex-QDs on the binding site of conA-AuNPs. The inhibition effect is concentration dependent. And using this NPs based RET system researcher can detect the deglycosylation and neoglycosylation states of protein using Avidin, NeutrAvidinTM, Bovine Serum Albumin (BSA) and diversely neoglycosylated BSA. Additionally we applied this system to discriminate the real glycoprotein products, recombinant glucose oxidase (rGOD) from yeas and fungus with varied branch types of mannosylation.

DNA Hybridization

DNA extracted from Allium Cepa (white onion) has been exposed to a variety of external thermodynamic parameters including: elevated temperature; low molar concentration salt solution; and very high simulated gravity. Results show that the actions of a dilute NaCl solution and of elevated temperature (to 65C) tend to unwind the double helix. This process is useful in nano-fabricating DNA polyhedra, particularly tetrahedra, in order to contain a trapped protein for microbiological applications such as anti-viral medication. Researcher show how the cellular structure of DNA changes its character and topology in response to extreme hyper gravity.

New Applications of Nanoparticles in Cardiovascular Imaging

Recently, four areas have emerged in cardiovascular imaging:

1. Targeted therapeutics: delivering drugs where they are needed;
2. Tissue engineering to replace defective valves, damaged heart muscle, clogged blood vessels;
3. Molecular imaging: using "smart" imaging agents in targeted therapeutics and molecular imaging;
4. Biosensors and myocardial diagnostics.

Several approaches of nanoparticles i.e. dendrimers, liposomes, polymer delivery molecules, cantilevers, nanoscaffolds, nanofibers are potential candidates in cardiac visualization. The extracellular matrix plays significant role by chemokines, cytokines and growth factors. The limitations of these emerging techniques and a new possibility of MRI visualization of mice cardiac atheroma by superparamagnetic iron-oxide gadolinium-apoferritin(SPIOA), myoglobin (SPIOM), for targeted functional and molecular imaging of atherosclerosis. These emerging techniques provide opportunity of tracking functional and structural changes in myocardium and heart tissue.

Application Areas for Nanoscale Pharmaceutical and Medicines Clusters of atoms

Material/Technique	Properties	Applications	Time-scale (to Market Launch)
Diagnostics			
Nano-sized markers i.e. the attachment of nanoparticles to molecules of interest	Minute quantities of a substance can be detected, down to individual molecules	CE.g. detection of cancer cells to allow early treatment	
'Lab-on-a-chip' technologies	Miniaturisation and speeding up of the analytical process.	The creation of miniature, portable diagnostic laboratories for uses in the food, pharmaceutical and chemical industries; in disease prevention and control; and in environmental monitoring.	Although chips currently cost over £1250 (US$2085) each to make, within three years the costs should fall dramatically, making these tools widely available.
Quantum dots	Quantam dots can be tracked very precisely by when molecules are 'bar coded' by thier unique light spectrum.	Diagnosis	In early stage of development, but there is enough interest here for some commercialisation (e.g. Q-dot Inc.).

(Contd.)

(Contd.)

Material / Technique	Properties	Applications	Time-scale (to Market Launch)
Drug Delivery			
Nanoparticles in the range of 50-100 nm.	Larger particles cannot enter tumour pores while nano-particles can easily move into a tumour.	Cancer treatment.	?
Nanosizing in the range of 100-200 nm	Low solubility.	More effective treatment with existing drugs.	?
Polymers	These molecules can be engineered to a high degree of accuracy.	Nanobiological drug carrying devices.	?
Ligands on a nanoparticle surface	These molecules can be engineered to a high degree of accuracy.	The ligand target receptors can reco-gnise damaged tissue, attach to it and release a therapeutic drug.	?
Nanocapsules	Evading body's immune system whilst directing a therapeutic agent to the desired site.	A Buckyball-based AIDS treatment is just about to enter clinical trials	Early clinical

(Contd.)

(Contd.)

Material/Technique	Properties	Applications	Time-scale (to Market Launch)
Increased particle adhesion	Degree of localised drug retention increased.	Slow drug release.	?
Nanoporous materials	Evading body's immune system whilst directing a therapeutic agent to the desired site.	When coupled to sensors, drug-delivering implants could be developed.	Pre-clinical: an insulin-delivery system is being tested in mice.
'Pharmacy-on-a-chip'	Monitor conditions and act as an artificial means of regulating and maintaining the body's own hormonal balance.	E.g. Diabetes treatment.	More distant than 'lab-on-a-chip' technologies.
Sorting biomolecules	Nanopores and membranes are capable of sorting, for example, left- and right-handed versions of molecules.	Gene analysis and sequencing.	Current—?

(Contd.)

(Contd.)

Material/Technique	Properties	Applications	Time-scale (to Market Launch)
Tissue Regeneration, Growth and Repair			
Nanoengineered prosthetics	Increased miniaturisation; increased prosthetic strength and weight reduction; improved biocompatibility.	Retinal, auditory, spinal and cranial implants.	Most immediate will be external tissue grafts; dental and bone replacements; internal tissue implants (Miles and Jarvis, 2001).
Cellular manipulation (Miles and Jarvis, 2001).	Manipulation and coercion of cellular systems.	Persuasion of lost nerve tissue to grow; growth of body parts.	More distant: 5-7 years.

(B) NANOELECTRONICNS

Nanoelectronics is a nanotechnology relating to the miniaturisation of electronic devices. The drive to increase device density and operating speed has meant that electronic devices have entered the nanoscale, and new fabrication techniques, components and changed properties have to be considered.

Researcher explore the electrical properties of devices and circuits with dimensions ranging from several micrometer (one thousands of a millimeter) down to one nanometer. At these dimensions quantum phenomena start play a dominat role. Electrons become extended objects with a wave character. Just as ordinary waves (e.g. water waves), electrons can interfere forming standing wave patterns. Such interference patterns are a direct manifestation of quantum behavior which can be visualized, for example with scanning-probe microscopes or by measuring the electrical resistance as a function of an external control parameter, such as the magnetic field.

Quantum interference is the basis for the formation of most of our matter in daily life. The reason that atoms bind together to form molecules or solid-state matter, such as gold for example, routes on quantum physics. It is the cooperative constructive interference of many electrons that forms the internal glue which keeps the matter together.

Researcher explore these and other quantum effects in model structures ranging from lithographically defined devices down to single molecules.

The tremendous impact of information technology (IT) has become possible because of the progressive downscaling of integrated circuits and storage devices. IT is a key driver of todays information society which is rapidly penetrating into all corners of our daily life. Integrated circuits are also known as microelectronics. The term micro derives from microfabrication technology, which embraces all highly sophisticated techniques like optical- and electron-beam lithography, metallization, implantation and etching that allow to generate structures on the scale of one micrometer.

What is "Nano" going to change? The answer appears to be straightforward: microelectronics will gradually evolve into nanoelectronics. In fact, this has already happened as can be seen from the smallest feature size of todays integrated circuits which is below of one micrometer. There is no fundamental reason why the continuous size reduction should stop in the next couple of years. It is very likely that it will continue for the next decade. Although, each time the structure size is reduced, technical problems—often very demanding ones—arise, but conceptually not much happens. The electronics will be based on silicon and functions the way it functions today. The rule of the game is smaller, faster and better. One must not be an expert to immediately see that this "straightforward" downscaling must sooner or later stop. Charge carriers in semiconductors are due to doping. Even if researcher would make use of highly doped material alone, only by chance could we find a single dopant in a 10 nm-sized transistor. Naturally, we will be dealing with a very few numbers of charge carriers, if at all, and control of charge and electrical current on a single electron level will be required. Moreover, quantum phenomena will increasingly start to dominate the overall behaviour of such structures. Finally, tinny structures have a large surface-to-volume ratio which is deadly for conventional semiconductor devices. Taking these facts, it is for certain that new concepts need to be developed!

Two possible approaches for obtaining structures within the size-regime of 1-100 nanometers are generally discussed. The first one is the top-down approach based on lithography, which is currently used to fabricate integrated circuits. It has been highly sucessful until now, but lacks control on the single atom level. The second one is the bottom-up (or chemist's) approach in which complex structures are assembled from single atoms and molecules into supramolecular structures. This supramolecular chemistry embodies single atom precision, at least in principle. The disadvantage is that we have no idea yet on how to assemble these molecules into useful operational devices. Though these two possible routes were known, most researchers in the field of nanoelectronics enjoyed their own training in traditional top-down technologies and no paradigm change was in sight.

Nanoelectronics (the Future)

Nanoelecetronics has witnessed a shift towards molecular systems in recent years. Though the term molecular electronics is a pretty old one, it is only since very recent that single molecules have become the focus of interest. To a great extent this was triggered by research on carbon nanotubes. Before the carbon nanotubes entered the scene, molecular electronics was the science of organic polymers, their synthesis, processing and doping. With carbon naotubes, researcher finally have a model system at hand that is equally of interest for chemists, material scientists and physicists. While carbon nanotubes are supramolecular objects for a chemist, they are one-dimensional solids for a physicist. In the future, more of these supramolecular structure will be studied on a single molecule level. Companies like, for example Motorola, IBM and Hewlett Packard, are starting to take an active part in this development. Single supramolecular structures are invisaged as switches and storage media. As has been shown already with DNA molecules, the trend towards molecule will include biological macromolecules as well. The ability to manipulate and characterize single molecules is an important first step for the exploration of suitable molecular functions. A fully functional chip, however, requires the ability to assemble the moelcules with high prescission into a functional network.

Quantum computing and quantum communication is another development that has recently taken place. Until now, these computers only exist as proposals on paper, but the theory has already been pushed very far. We know today that a quantum computer will outperform any classical computer. We are now faced with the challenge to realize a demonstrator of such a machine, if possible one that can be scaled up using integration and mass fabrication. A new territory has been discovered for the science of micro- and nanoelectronics. Unlike classical computers, in which bits are either one or zero, quantum computers make use of coherent superpositions of "zero" and "one". In contrast to a classical "switch", the quantum switch can be both open and close, because it may be in a coherent superposition of the two possibilities. The quantum computer derives its power from this unambiguity. However, this peculiar feature is lost after some time by interaction with the environment. This so-called "dephasing" time determines the number of operations that can be performed by a quantum computer before it has to be "reset". We have to learn how to manipulate coherent superpositions in the solid and to understand the limiting factors leading to depasing.

Finally, there is a third direction in nanoelectronics which will receive more attention in the future. This new field is called "spintronics". Spintronics is concerned with electromagnetic effects in nanostructures and molecules caused by the quantized angular momentum (the spin) that is asscociated with all fundamental particles like, for example, the electron. The magnetic moment of a particle is directly proportional to its spin. Hence, if we learn to manipulate not only charge, but also spin on a single electron level, information may be stored and transported in the form of quantized units of magnetism in the future.

Plastic Electronics

There is a great scope for plastic eletronics because of flexibility, ease of processability and ease of fine tuning of the structures. However, this area is still in infancy because of the limited availability of semiconducting polymers. Polyaniline is easily available but is not stable to air or high temperatures and hence could not be used for commercial applications though it has been shown to exhibit transistor behavior. In case of PEDOT, though it is stable to air and elevated termperature, it can be used because it is not available in semiconducting form essential for nanoelectronics. The Baytron P is the only processable PEDOT materials, but it is in conducting form and can not be isolated in semiconducting form. Therefore, in order to get the reduced semiconducting PEDOT, one has to rely on true soluble PEDOT derivatives.

Processes for Nanoelectronics

For making nanosized devices, processing parameters of conventional processes and novel processes are necessary. Modifications to conventional processes include low thermal budget and low energy implant and plasma processes to maintain sharp doping profiles and have low process damage. Novel processes include those necessary to deposit/remove new materials on silicon wafers or a modification to a conventional process that creates a deposited layer with novel properties. An important new class of processes are those concerned with locating nanodevices or parts of devices created via synthetic routes on to electrical leads, such as, for example, locating a single 100nm carbon nanotubule on an electrode pair. For the materials we are working with we would need to look at the following issuses to make the transition from researching the above materials to making devices. Researchers will look at following issues.

- Low thermal budget processes for nanoscale devices (CMOSplus + siliconplus).
- Processes for incorporating non-conventional materials in integrated circuits—the materials we shall look at are the semiconducing materials in the list above plus metals and dielectrics (CMOSplus + siliconplus).
- Processes for creating interconnects to nanosized devices (siliconplus).

(C) NANOSENSING

Nanosensing uses small particles such as quantum dots to detect biological molecules, and indicate their presence either by the light emitted in luminescence or the light scattered. The use of specifically funtioned nanoparticles for the detection and identification of biomolecules is expected to revolutionise biosensing and medical diagnostics.

Conducting Polymer Based Sensors

In pursuit of the holy grail of molecular electronics, molecular materials have a fundamental advantage. They have the necessary electronic structure at the molecule level as well as the possibility of tuning it. Some of them have performed well in μm sized devices and can truly be built in a bottoms up manner. Conducting polymers, as a class, have come out well ahead in this race and have the potential, along with carbon nano structures, of producing working devices of nanometer size.

1. Electronic Tongue
2. Electronic Nose
3. Electronic Tongue

Such structures have a great potential in developing low-cost diagnostic kits for health-care. Another area of potential application is in electromechanical devices where conducting polymers have a great advantage due to their unique electrochemical properties. Structures like cantilevers and diaphragms can be built by the bottoms up approach. such studies will provide a window to the fundamental molecular mechanisms involved in the transduction process and a direct measure of the inter-molecular forces. The expansion/contraction of the polymer film is the result of processes of ion insertion/ejection, solvation/desolvation, and coiling/uncoiling of polymer chains. Forces generated/exerted by these interactions are rarely accessible in such a direct manner as they are here. With nanometer size structures the signal/noise ratio of such effects will become more favorable and one can hope to see differences with change in ionic radii, polarity of solvent, and polymer structure. This could lead to a way to estimate the magnitude of forces involved in weak interactions.

Ceramic actuators and gas sensors are being developed and a nano-structured composite material will have advantages in terms of sensitivity and ease of fabrication.

Electronic Tongue

Lab on a Chip: Biomedical Sensors: We have demonstrated and published our results sometime back that PEDOT can be used for immunosensing and also for DNA sensing. These biomedical applications have huge commercial potential. However, we could not take this concept to next level because the sensors were made from electropolymerization route. This means one can make one sensor at a time and hence not viable for bulk production. However, we have now developed the soluble PEDOT materials, this means that we can now use the direct printing method for the fabrication of DNA and immuno and other biomedical sensors as printing is suitable for scale up.

Electronic Nose

Nanofibers by electrospinning: Electrospinning has been projected as a method of choice for the future materials based on electroactive nanofibers. These nanofibers are considered to be the materials which will solve some of the crucial problems in the field of environment, energy and military (gas sensing for chemical and biological warfare agents) applications. The conducting polymers have been shown to be the best materials for gas sensing. The only bottleneck in this area is the availability of the soluble conducting polymers wherein the properties can be fine tuned with ease. Our materials fulfill these criteria?s of solubility and easy fine tuning of the final structures and hence are of importance for these applications.

Let us understand Olfactory model.

- Traditional "one-sensor-one-analyte" sensing is not effective for multiple target agents in uncontrolled environments
- Olfaction is a highly sensitive, selective, fast and reliable sensing system that works well in "noisy environments"
- Some examples of what the nose can do:
  - Human females can detect minute differences in male immunotype by smell
  - A moth can smell a single molecule of the moth pheromone 'bombykol'
  - A bloodhound can pick up a 24 hour-old trail and identify the person
  - Rats need only one sniff (< 200 ms)
  - to accurately discriminate odors*

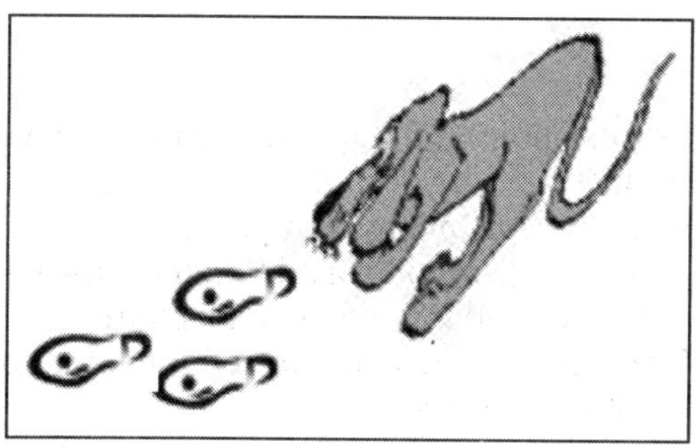

Major Steps in Olfactor Processing

- Odor molecules make contact with the olfactory receptor (OR) neurons
- The signals from the OR neurons are sent to the olfactory bulb for preprocessing
- From the olfactory bulb, the signals travel to the olfactory cortex and to other regions of the brain.

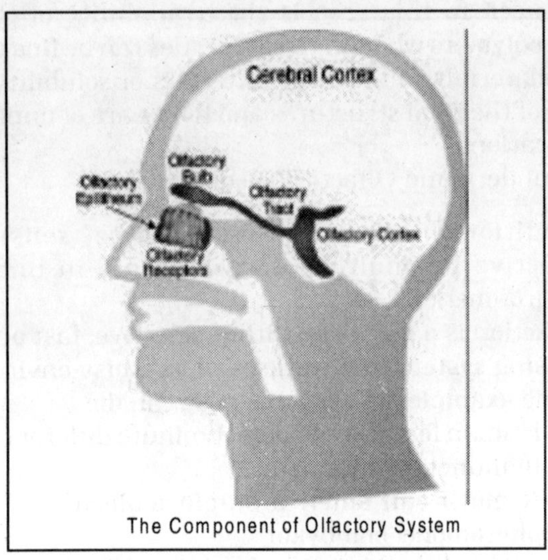

The Component of Olfactory System

The Transduction Method

- Odor molecule binds with a G protein-coupled receptor on the surface of the cilium

- Odorant-receptor binding activates the Golf protein inside the cilium
- Golf activates AC (adenylyl cyclase).
- AC catalyzes conversion of ATP (adenosine triphosphate) to cAMP (cyclic adenine monophosphate)
- cAMP travels throughout the cell, opening ion channels in the cell membrane
- Flow of positive ions in and negative ions out reduces the potential across the membrane
- If depolarization reaches threshold, an action potential is generated.

Early Processing in the Epithelium

In the Epithelium:

- Each OR neuron expresses a single receptor type

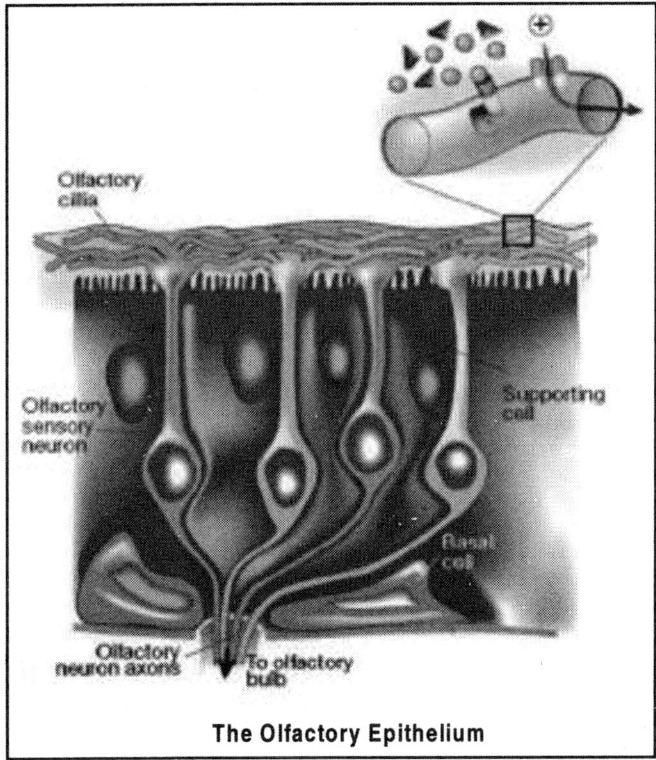

The Olfactory Epithelium

- Each receptor responds to a range of odorants with similar structures
- Incoming odorants may activate several different types of OR neurons
- All OR neurons of the same type send their axons to the same glomerulus in the olfactory bulb.

Signal Flow to the Olfactory Cortex

Olfactory Processing

- In the bulb, inputs from different ORs are segregated by type
- Activated glomeruli in the bulb provide a stereotyped map or "code" for the odorant

- In the olfactory cortex, individual cortical neurons may receive input from multiple ORs
- Signals reach multiple areas of the brain, resulting in another map of sensory information.

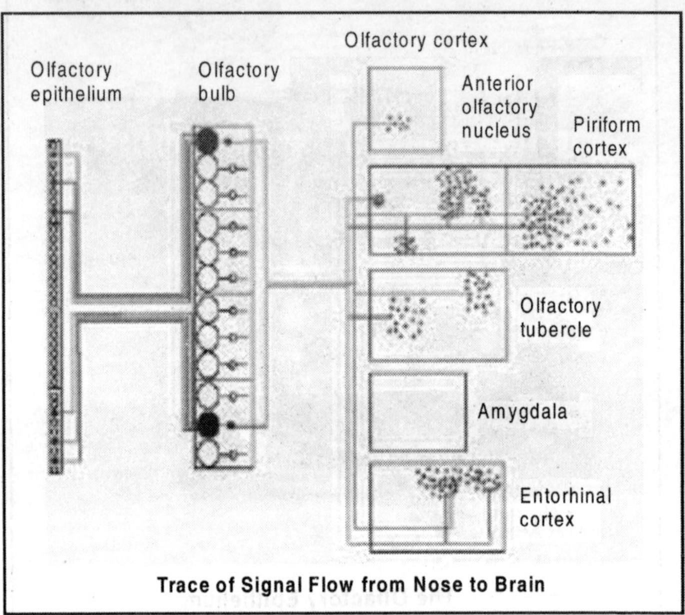

Pattern Analysis for Artificial-Nose

- Signal Preprocessing, e.g. compensation for sensor drift, vector normalization
- Dimensionality Reduction, using Principal Component Analysis (PCA), Fisher's Linear Discriminant Analysis (LDA), etc.
- Classification, e.g. artificial neural networks, parametric methods, partial least squares

Toxonomy of Nose-Like Sensors

The Kamina: A Micro-Scale Tin-Oxide Artificial Nose

The KAMINA detects gases such as ammonia, benzene, carbon monoxide and formaldehyde.

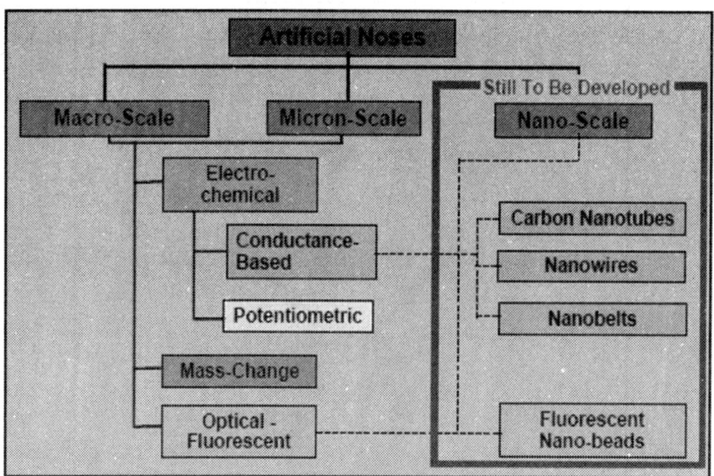

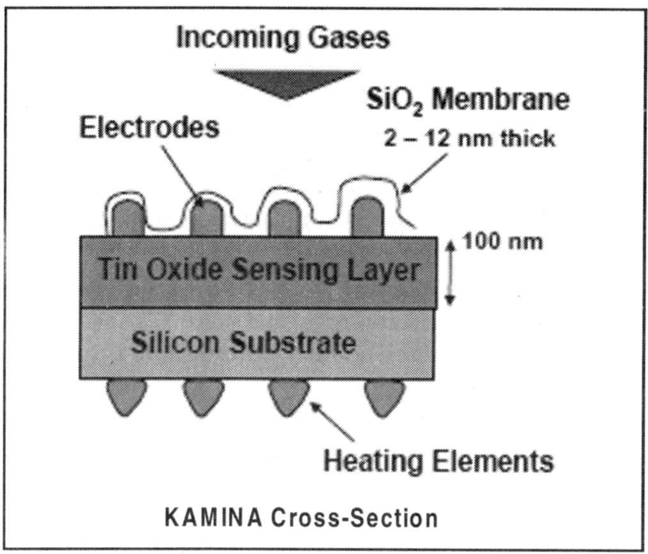

Artificial Noses

Based on High-Density Optical Array

- Thousands of microsphere sensors are dispersed across the face of an etched optical imaging fiber

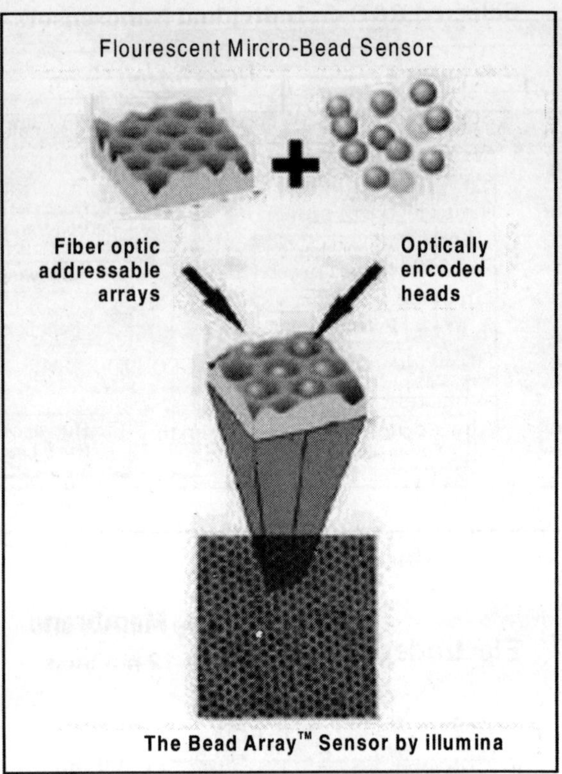

- Interaction of odorant with fluorescent dyes yields a frequency shift in the returned signal
- Each bead is independently addressable through its own light channel
- Fiberoptic artificial noses have been demonstrated to detect land-mines

A Smart-Chip Chemical Microsensor

- Incorporates three different transducers, each with sensitive polymeric layers:
  - ➢ Mass-sensitive
  - ➢ Capacitive
  - ➢ Calorimetric
- On-chip control and monitoring of the sensor functions
- Commercial CMOS fab process: chip is 7x7 mm

Selected R&D on Individual Nanosensors

Type	Description
Carbon Nanotube	Semiconducting-SWNT sense NH_3, NO_2 and H_2 by undergoing a conductance change
Carbon Nanotube	SWNT show high sensitivity to oxygen or air, changing their electrical properties
Nanowire	Silicon nanowires in solution sense streptavidin, biotin, etc. with conductance changes
Nanowire	Tin oxide nanowires sense O_2 and CO with conductance changes
Nanobelt	Tin oxide nanobelts sense CO, NO_2 with a change in conductance
Nanobead	'Almost' nanometer-scale sensing beads used in optical sensors

Silicon Nanowires (NW) as Chem/Bio Sensors in Solution

- Lieber's group at Harvard tested silicon NWs modified differently for each of four sensing operations

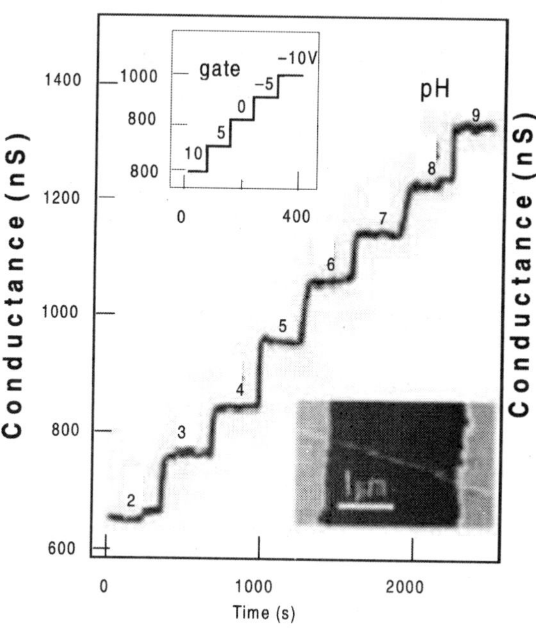

- Discrete changes in pH levels from 2 to 9 caused stepwise increases in conductance
- Biotin-modified NWs show a conductance increase in the presence of streptavidin, but the process is not reversible
- Monoclonal antibiotin (m-antibiotin) binds reversibly with biotin, decreasing conductance
- Calmodulin-modified NWs reversibly sensed Ca^{2+} with a conduction decrease

Semiconducting Oxide Nanobelts as Gas Sensors

Wang at Georgia Tech and Sberveglieri at U. Brescia, Italy, demonstrated that single-crystalline SnO_2 nanobelts undergo a change in conductance in the presence of CO, NO_2, and ethanol

Towards Nanometer-scale Sensing Beads

- Sensing beads, about 300nm in diameter, are deposited in microwells on the surface of optical fibers

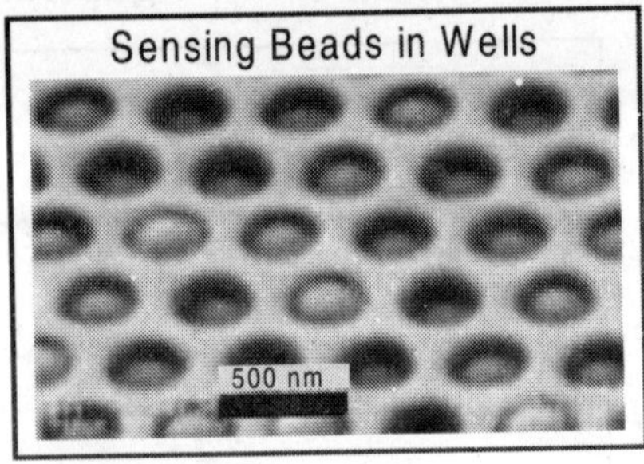

- Binding of a target agent produces a shift in wavelength of returned light
- These sensing systems, developed at Tufts Univ., detect nitroaromatic compounds similar to low-level explosives

Nose-Like Nanosensing: Where Are We Now?

- Micron-scale sensing elements have been developed and incorporated into small artificial noses
- Nanometer-scale devices demonstrated to function as gas sensors

However, at this time, no nose-like sensing systems integrated on the nanometer-scale have been developed
- Some of the Challenges Remaining:
    - Complete characterization of sensing behavior of nano-scale devices
    - Precise placement versus random distribution of nanosensors ?
    - Nanosensor system/microprocessor integration
- Useful Next Steps: Development and Integration of
    - Small number of differentially sensitized nanosensors
    - Microprocessor/processing algorithm into operational nose-like nanometer-scale sensing system.

Building upon Nanomemory Systems for Nose-like Sensing

- This cross-bar array of nanowires is a nano-sensing system: each junction of 2 crossed nanowires forms a memory bit

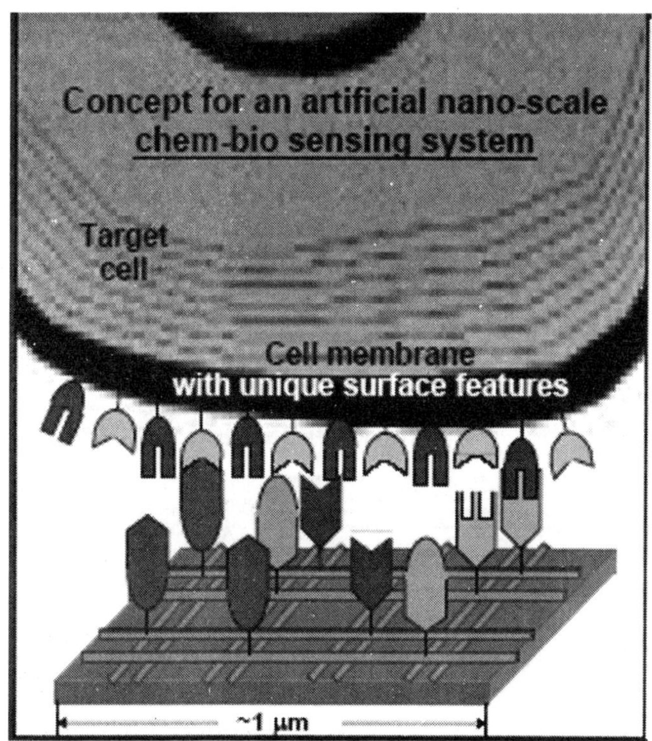

- The junctions of the nanowires are decorated with binding agents
- Binding of a target agent at a memory location causes the memory bit to change from "0" to "1"
- The set of memory bits reporting a change in status provide an identifier for the target agent

Nose-like Nanowire Sensing System

- Core of sensing element is SnO_2 nanowire, as demonstrated by Moskovits.
- Modify by coating the wires with a layer of SiO_2
- As demonstrated by Goschnick, differential thicknesses of SiO_2 coatings over SnO_2 cause differences in conductivity patterns
- A system of such nanowires, combined with a heating element, should behave as an electronic "nano-nose"

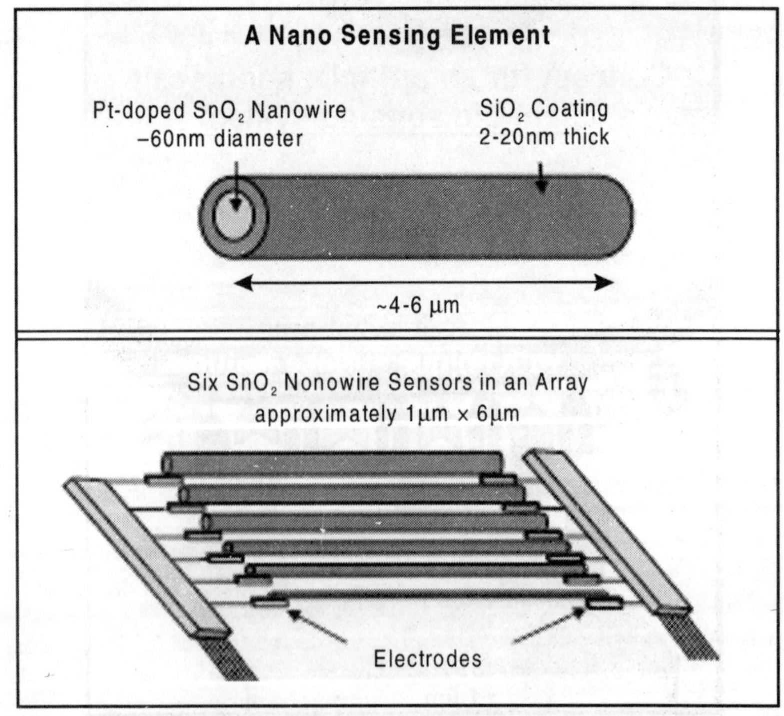

(D) NANOMAGNETICS

Nanomagnetics is a nanotechnology relating to the miniaturisation of magnetic materials. This area is developing in conjuction with the drive for more powerful and higher capacity computer-driven devices.

Magnetic nano-elements have great potential for technological applications, like magnetic disks and Magnetic sensors. The formation of domains and the specific domain structure plays a crucial role in all applications.

Computer hard drive capacity could be increased a hundredfold by using a common protein to fabricate nano-scale magnetic particles.

It uses the protein apoferritin, the main molecule in which iron is stored in the body, to create a material consisting of magnetic particles each just a few nanometres in diameter.

Each particle can store a bit of information and together they can be packed onto a disk drive at much greater density than is possible using existing hard disk manufacturing methods.

It sounds simple and relatively it is. It provides the hard drive industry with a roadmap to go from pretty near the nano-scale to the ultimate extent of magnetic recordings. The material is still in development and could be available between six to 10 years from now.

Hollow Core

Apoferritin proteins are spherical and measure 12 nanometres in diameter, with an eight nanometre hollow core. When iron fills the core the combination is called ferritin.

The process developed by Nanomagnetics involves treatment with an acid solution to remove the iron core from ferritin. This is followed with a second solution that fills the cores with a magnetic cobalt-platinum alloy.

The resulting solution, dubbed DataInk, is then sprayed onto the surface of a hard disk and treated with heat. This alters the crystalline structure of each particle encouraging them to self-assemble into a tightly packed single layer. Heating also turns the protein's outer shell into carbon.

Currently, about 450 gigabits of data can be squeezed onto a square centimetre of disk. Manufacturers believe this could eventually be improved to a maximum of 3000 gigabits per square centimetre.

But Mayes says Nanomagnetics new material could allow 5000 gigabits per square centimetre to be stored. He adds that the means of writing and reading data to disks coated with DataInk are already on the drawing board.

Up or Down

Seagate is working on a nano-magnetic material that is fabricated chemically, rather than using proteins.

The most significant problem scientists are facing is that particles are not automatically oriented in a uniform way, that is up or down, so that they can store a magnetic charge.

But the problem can overcome by the application of a magnetic field during the film formation and heating process.

Today, about 90 percent of all new data created reside on magnetic media, primarily hard disk drives. Every day new applications are incorporating hard disk drives to allow for the rapid storage and retrieval of important data, driving more than $30 billion in revenue in 2005. Over the past 50 years, areal density, the measurement of how many data bits can be stored on an inch of disk space, has increased over 50 million times. At the same time, drives have gone from filling a room to fitting in a cell phone. New technologies such as perpendicular recording and tunneling magnetoresistive heads are continuing the acceleration of storage density today. The hard disk drive is an amazing collection of technologies requiring innovative engineers and scientists from many diverse backgrounds. The challenges that are present in extending hard disk drive technology into the future through Heat Assisted Magnetic Recording (HAMR) and discrete patterned media will prove exciting for the next generation of storage technology.

Tunnel Magnetoresistive

TMR effect is a magnetoresistive (MR) effect that occurs in a tunnel junction in which a thin insulator is held between ferromagnets (with a ferromagnetic layer on both sides of the insulating layer).

Similar to other MR effects, such as anisotropic magnetoresistive (AMR) effect and giant magnetoresistive (GMR) effect, its conductivity is spin-dependent. The first report on the TMR effect was made by Julliere in 1975. Thereafter, several groups reported on it, but because of the small rate of change (less than 1%) in MR at room temperature, the reports did not receive significant attention.

In 1991, the prizewinners discovered TMR effect of 2.7% at room temperature; in 1994, they discovered a giant TMR effect of 18% at room temperature with a $Fe/Al_2O_3/Fe$ junction. Since then, the prizewinners have led the development in Japan of magnetic playback heads and solid-state magnetic memory using TMR elements. The specific issues they encountered included : (1) optimizing junction electrode materials, (2) making junction tunnel resistance low and (3) improving TMR effect and thermal stability through thermal treatment. The technology for these and related achievements included (i) technology to manufacture and evaluate uniform and thin insulating layers and (ii) planarizing technology for interfaces between magnets and insulators and technology to evaluate interface condition. For the former, an ICP plasma oxidation method and Conducting Atomic Force Microscope were adopted for manufacturing thin insulating layers (low-resistance tunnel junctions); for the latter, Inelastic Electron Tunneling Spectroscopy was adopted for interface evaluation, enabling successful manufacture of quality tunnel junctions. As a result, high TMR ratios of 50% at room temperature and 77% at 4.2 K could be obtained:.

Recently, tunnel junctions with MgO barriers and tunnel junctions that use Heusler alloys for electrodes have been the subject of attention; the prizewinners are world-leading researchers on tunnel junctions using Heusler alloys.

Significance of Achievement

The first achievement of this research to be cited is its contribution to the development and practical application of high-density HDDs in the 100 Gbit/inch^2 class, along with its contribution to the development of MRAM, a nonvolatile memory. Although MRAM has yet to reach the practical application stage, samples are being shipped and R&D is being conducted around the world, with expectations increasing daily. In addition to contributing to these industry developments, this research also has major academic significance, as it contributes to the establishment of the discipline of spin electronics. Henceforth, through rapid development of the spin electronics field, zero-standby-power computers with energy-saving memory that will not be lost even in the event of fires or disasters will become possible, thus contributing to the actualization of an environment-friendly society.

Reasons for Selection

To boost the recording density of the hard disk drive (HDD), which is the storage medium for computers, in addition to its recording performance to magnetic discs, improvement of their reading performance is extremely effective. Making high-performance magnetic heads for this purpose can thus be compared to the history by which high recording density was developed for HDDs. A review of this history to the present reveals that first, magnetic heads utilizing AMR effect emerged for achieving high recording density, followed by magnetic heads utilizing GMR effect. However, improvement of recording density knows no stopping. Eventually, people came to expect recording density exceeding 100Gbit/inch^2, which was difficult to achieve with magnetic heads utilizing GMR effect.

The development of magnetic heads utilizing the TMR effect has been active in recent years, as they are considered to be superior to heads that utilize the GMR effect. Thus, the possibility has arisen that all GMR heads will be replaced with TMR heads. The TMR effect is a magnetoresistive effect that occurs in a tunnel junction in which a thin insulator is held between ferromagnets. The first report on the TMR effect, at a low temperature (4.2K) was made in 1975 by Julliere of University of Rennes, France. Although there have been numerous reports since then, they have not received much attention, owing to their low MR ratio at room temperature: under 1%.

Terunobu Miyazaki has been researching MR effect since around 1985. Inspired by a report on GMR involving an artificial metal lattice in 1988, he undertook research on the TMR effect, which involves the same spin-dependent conductivity. In 1991, Miyazaki discovered TMR effect of 2.7% at room temperature, which led to his discovery of a giant TMR effect with an MR ratio of 18% with a Fe/Al$_2$O$_3$/Fe junction in 1994. Since then, he has led Japan's development of TMR magnetic heads. During this time, he has also focused on developing important technology geared to practical application, such as by optimizing electrode and insulation materials, making tunnel resistance low and stabilizing the TMR effect. His research papers have been cited 281 times throughout the world, and from the fact that his announcements have led to sudden increases in patent applications pertaining to TMR, as well, the major impact that his TMR-related research has had on the world is palpable. At present, the US company Seagate Technology, Inc. has announced its full-scale adoption of a 2.5-inch HDD using TMR heads, and TDK has announced that in 2006, it will be incorporating TMR in perpendicular magnetic

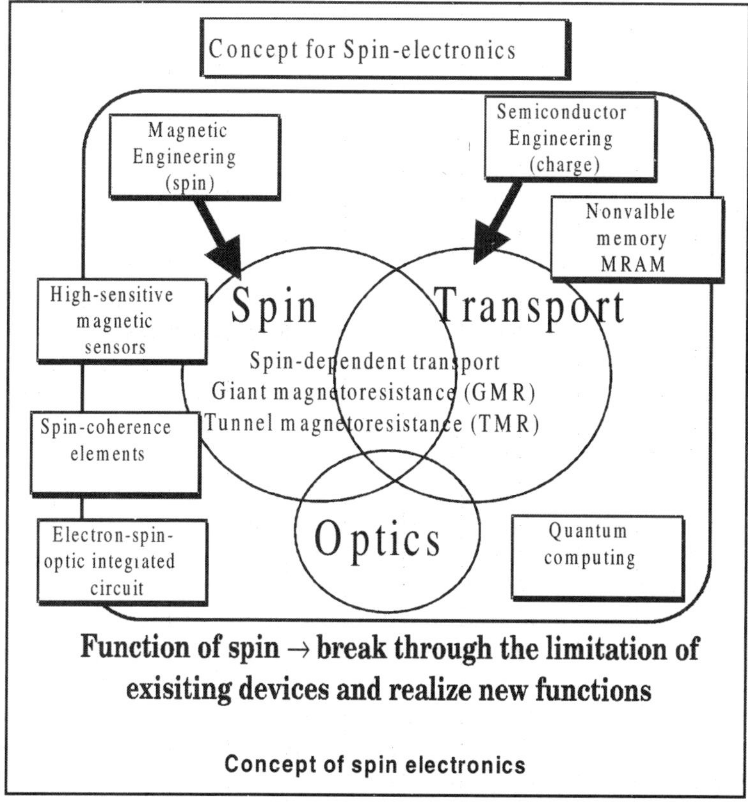

Concept of spin electronics

recording for mass production. Hitachi and Fujitsu have also reported their adoption of TMR heads. Even with the various contrivances each company has made for practical application, Miyazaki's achievement in constructing the basis for them is immense. In addition, TMR technology will be the source for the development of MRAM, of which much is expected as a memory element for the future.

As the demand for storage has increased dramatically over time, technologists nave worked toward increasing the amount of information that can be stored onto disc drives. By increasing the areal density—or the amount of information that can be placed within a given area on a disc drive—technologists in fact have been able to deliver densities in excess of 100 percent annually over the course of the last several years. A key end-result or benefit of this dramatic areal density curve is that disc drive manufacturers have also been able to drive down the cost of the disc drives themselves because they can offer higher capacity disc drives using fewer platters, heads, and mechanical parts.

For the past 40 years, longitudinal recording has been used to record information on a disc drive. In longitudinal recording, the magnetization in the bits on a disc is flipped between lying parallel and anti-parallel to the direction in which the head is moving relative to the disc. Each bit itself is made up of approximately 100 magnetic grains.

Increasing areal densities to allow greater capacities is no small task. Today it is becoming more challenging to increase areal densities in longitudinal recording—previous growth rates have typically been well over 100 percent annually, while the average is approximately 60 percent currently.

To increase areal densities in longitudinal recording, as well as increase overall storage capacity, the data bits on a disc must be made smaller and put closer together. However, there are limits to how small the bits can be made. If the bit becomes too small, the magnetic energy holding the bit in place may also become so small that thermal energy can cause it to demagnetize. This phenomenon is known as superparamagnetism. To avoid superparamagnetic effects, disc media manufacturers have been increasing the coercivity (the "field" required to write a bit) of the media. However, the fields that can be applied are limited by the magnetic materials from which the write head is made.

Perpendicular recording will occur sometime between 100 and 200 gigabits per square inch areal density. In perpendicular recording, the magnetization of the disc—instead of lying in the disc's plane as it does in longitudinal recording—stands on-end perpendicular to the plane of the disc. The bits are then represented as regions of upward or downward directed magnetization.

The ideal perpendicular recording media will have an M-H loop with unity squareness (S) to avoid excess noise in the DC saturated state, a negative nucleation field in excess of the fields produced under the return pole of the recording head, a high anisotropy to provide thermal stability and a small average grain size and small intra-granular exchange to reduce transition jitter. These requirements to varying degrees have been met by two material systems used in recent demonstrations of perpendicular recording, CoCrPt alloys and Co/Pd (Co/Pt) multi-layers

Breakthrough Nanotechnology Will Bring 100 Terabyte 3.5-inch Digital Data Storage Disks.

Michael invented and patented the world's first and only concept for non-contact UV photon induced electric field poling of ferroelectric non-linear photonic bandgap crystals, which offers the possibility of controlling and manipulating light within a UV/Deep Blue frequency of 1 nm to 400 nm.

It took him 14 years to find a practical conceptualization that would work to advance the storage industry; 3D Volume Holographic Optical Storage Nanotechnology, for which Michael holds the patents. He was invited to present this fascinating discovery to the National Science Foundation in February 2004.

This invention and patents on a technique for changing matter at the molecular level is one of the World's only new enabling technologies, having many hundreds of electro-optic applications.

Atomic Holographic Nanotechnology will allow for the first time a functional method for programmable molecular lenses that will allow incoming light to be rejected, modified internally, or allowed to pass unaltered through a transparent lens known as disk, tape, card, drum, film, etc.

By being able to program optical lenses, many applications based on light and color can be developed, such as holographic storage, bio-terror detection devices, optical electronics, security products, and hundreds of other products never seen before on the world's markets.

The small size of ferroelectric transparent structures makes it possible to fabricate nano-optical devices, such as volume holographic storage, having both positive and negative index of refraction that will allow molecular particles of an atomic size to be modified, controlled, and changed to perform a specific function, desired task, used for low cost accurate chemical/biological matter detection, and reprogrammed to accept new non-volatile data and molecular functions.

Potential Advantages of Using Spin Transfer Instead of Magnetic

Fields to Make RAM

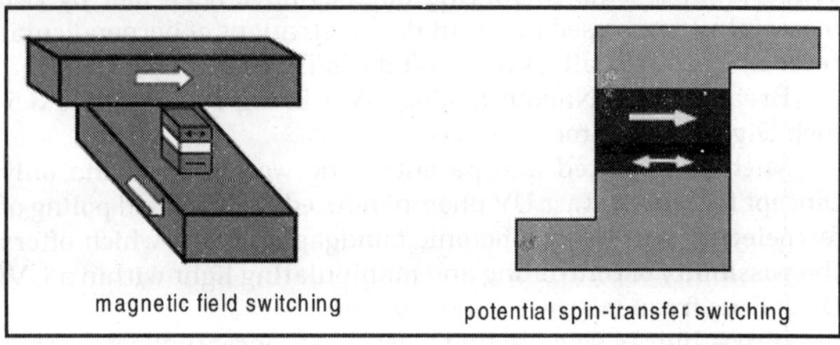

- Spin transfer gives stronger torques per unit current than for magnetic fields, in devices smaller than about 250 nm.
- Spin transfer gives short-range forces. No "half-current" problem.
- Excellent scaling to small sizes—switching currents can be minimized while maintaining magnetic anisotropy barriers for thermal stability
- Spin-transfer allows simpler device geometries, with less demanding device tolerances.

Fundamentally New Tool to Control and Manipulate Magnetic

nanostructures

- current induced switching
- coherent dynamic precession

Technology

- Integrable with semiconductor technology
  - ➤ microwave source, mixer, transmitter, signal processing
- 'SMT-enabled' magnetic devices—MRAM, spintronics
- 'SMT-disabled' magnetic devices—future hard disk read heads

New Physics

- Insight into spin transport and interactions
- Fundamentally new types of magnetic excitations
- Most of the theories are still untested.

Spin Transfer Induces Precessional Oscillations

in plane—simple linear red shift with current
out of plane—blue shift with current
intermediate angles—complicated mode structure
Qualitative agreement with single domain modeling

Application of Nanomagnetic Particles in Hyperthermia Cancer

Treatment

Scientists are focusing on developing uniform particles with a Curie temperature that is similar to the therapeutic one for the purpose of cancer treatment. Such particles will self-regulate the temperature of the tumor during Magnetic Hyperthermia (MH), thus avoiding the use of temperature controls. Magnetic hyperthermia is based on a defined transfer of power onto magnetic nanoparticles in an alternate magnetic field that is determined by the frequency, magnetic field strength, materials and the size of particles, which results in local generation of heat. This heat will either destroy the tumor cells directly or result in a synergic reinforcement of radiation efficacy, depending on the equilibrium temperature set in the tumor tissue. MH involves the local deposition of tumor cell specific magnetic nanoparticles and an external alternating current magnetic field applicator system. Nanoferrite particles can be easily applied interstitially for minimal invasive application. In this study more uniform nanoparticles have been manufactured and they were tested for their Curie temperature for self regulated MH.

BCC report identifies current, emerging, and futuristic applications for nanomagnetic materials and devices, and evaluates actual and potential markets by application and type. Some nanomagnetic products have been on the market for years, or even decades, whereas others have yet to reach commercialization.

7

RECENT DEVELOPMENT

Nanotech Water Desalination Membrane

Researchers at the UCLA Henry Samueli School of Engineering and Applied Science today announced they have developed a new reverse osmosis (RO) membrane that promises to reduce the cost of seawater desalination and wastewater reclamation.

Reverse osmosis (RO) desalination uses extremely high pressure to force saline or polluted waters through the pores of a semi-permeable membrane. Water molecules under pressure pass through these pores, but salt ions and other impurities cannot, resulting in highly purified water.

The new membrane uses a uniquely cross-linked matrix of polymers and engineered nanoparticles designed to draw in water ions but repel nearly all contaminants. These new membranes are structured at the nanoscale to create molecular tunnels through which water flows more easily than contaminants.

New Biomedical Device Uses Nanotechnology to Monitor Hip Implant Healing

It is so small, you can barely see it, but a microsensor created by University of Alberta engineers may soon make a huge difference in the lives of people recovering from hip replacement surgery. The U of A research team has invented a self-powered wireless microsensor for monitoring the bone healing process after surgery—it is so tiny it can fit onto the tip of a pen.

"This microsensor not only reduces post-operation recovery time, it will also help reduce the wait time for patients needing artificial joint implants

During the healing process that follows joint replacement, bone grows and attaches to the pores on the surface of the implant creating greater fixation and stability of the joint. This process is known as osseointegration.

Using nanotechnology, the researchers built a device that measures and compares the relative osseointegration of a hip implant over time. The microsensor will be able to monitor the progression of the biological fixation between bone tissue and the implant. The sensor is permanently implanted with the joint and is powered kinetically—it uses the natural movement of the patient's body as its power source. When it isn't being used, it stays dormant until a doctor asks it to start transmitting data.

Careful monitoring of how the patient is healing will help patients recover as quickly as possible and resume normal activities with less chance of over stressing the fracture during recovery and rehabilitation. It also allows the surgeon to more accurately decide when it is safe to send patients home from the hospital with their new implants.

NIST Laser-Based Method Cleans Up Grubby Nanotubes

Before carbon nanotubes can fulfill their promise as ultrastrong fibers, electrical wires in molecular devices, or hydrogen storage components for fuel cells, better methods are needed for purifying raw nanotube materials. Researchers at the National Institute of Standards and Technology (NIST) and the National Renewable Energy Laboratory (NREL, Golden, Colo.), have taken a step toward this goal by demonstrating a simple method of cleaning nanotubes by zapping them with carefully calibrated laser pulses.The researchers developed the new method while looking for quantitative methods for evaluating laser damage to nanotube coatings for next-generation NIST standards for optical power measurements. The responsivity of a prototype NIST standard increased 5 percent after the nanotube coating was cleaned.

Chemists Create 'Nanorobotic' Arm to Operate within DNA Sequence

New York University chemistry professor Nadrian C. Seeman and his graduate student Baoquan Ding have developed a DNA cassette through which a nanomechanical device can be inserted and function within a DNA array, allowing for the motion of a nanorobotic arm.

Butterfly Wings are Templates for Photonic Structures

By replicating the complex micron- and nanometer-scale photonic structures that help give butterfly wings their color, researchers have demonstrated a new technique that uses biotemplates for fabricating nanoscale structures that could serve as optical waveguides, optical splitters and other building blocks of photonic integrated circuits

Boffins Seveloping Nanoparticle Implant to Monitor Tumour Growth

Boffins at the Massachusetts Institute of Technology (MIT) are developing a tiny implant that they say could help doctors rapidly monitor the growth of tumours, as well as the progress of chemotherapy in cancer patients, in the future.

The implant contains nanoparticles that can be designed to test for different substances, including metabolites such as glucose and oxygen that are associated with tumour growth.

The boffins say that the implant will also be able to track the effects of cancer drugs, for once inside a patient, it could reveal how much of a certain cancer drug has reached the tumour, thus helping doctors determine whether a treatment is working in a particular patient.

The team is being led by Michael Cima, the Sumitomo Electric Industries Professor of Engineering in the Department of Materials Science and Engineering.

And while nanoparticles have been used before, the implant being created by the MIT researchers is unique as for the first time boffins have encased the nanoparticles in a silicone delivery device, allowing them to remain in patients' bodies for an extended period of time.

The device can be implanted directly into a tumour, allowing researchers to get a more direct look at what is happening in the tumour over time.

The new technique, known as implanted magnetic sensing, makes use of detection nanoparticles composed of iron oxide and coated with a sugar called dextran. Antibodies specific to the target molecules are attached to the surface of the particles. When the target molecules are present, they bind to the particles and cause them to clump together. That clumping can be detected by MRI (magnetic resonance imaging).

In addition to monitoring the presence of chemotherapy drugs, the device could also be used to check whether a tumour is growing or shrinking, or whether it has spread to other locations, by sensing the amount and location of tumour markers.

The researchers are now preparing a paper on the work and have presented their findings at recent meetings of the European Cancer Society and the American Institute of Chemical Engineers

Samsung Electronics Develops 50-Nano Memory Chip

The world's largest memory chip maker said that it will mass produce the new memory, mostly used in personal computers, in the first quarter of next year. The company expected its global market to grow to US$55 billion by 2011.

Currently, chip making companies are using 80-nano technology in producing DRAMs.

Samsung said the latest manufacturing process, the first of its kind in the world, will enhance productivity by 200 percent compared with the existing one.

The company has been leading global efforts to develop enhanced chipmaking processes since 2000 when it unveiled the world's first memory chips based on 150-nano technology. It was also the first to develop DRAM chips using 60-nano technology last year.

Scanning Photoionization Microscopy

SPIM combines the high spatial resolution of optical microscopy with the high sensitivity to subtle electrical activity made possible by detecting the low-energy electrons emitted by a material as it is illuminated with laser pulses. The technique potentially could be used to make pictures of both electronic and physical patterns in devices such as nanostructured transistors or electrode sensors, or to identify chemicals or even elements in such structures.

Apparatus includes a moving optical microscopy stage in a vacuum, an ultrafast near-ultraviolet laser beam that provides sufficient peak power to inject two photons (particles of light) into a metal at virtually the same time, and equipment for measuring the numbers and energy of electrons ejected from the material. By comparing SPIM images of nanostructured gold films to scans using atomic force microscopy, which profiles surface topology, the researchers confirmed the correlations and physical mapping accuracy of the new technique. They also determined that lines in SPIM images correspond to spikes in electron energy, or current, and that contrast depends on the depth of electrons escaping from the metal as well as variations in material thickness.

Work is continuing to further develop the method, which may be able to make chemically specific images, for example, if the lasers are tuned to different colors to affect only one type of molecule at a time.

High Efficiency Solar Cell

Luna nano Works is nearing commercialization of a novel version of buckyballs—soccerball-shaped carbon molecules—that the company says could improve magnetic resonance imaging (MRI) and lead to high-efficiency solar cells. Each buckyball is made of 80 carbon atoms with metal-nitride clusters trapped inside, creating a nanomaterial with novel electronic, optical, and magnetic properties.

Scientists typically create buckyballs—hollow spheres made of 60 carbon atoms are the most common kind—by striking an electric arc between two graphite electrodes.

These are the first buckyballs enclosing highly unstable metal-nitride molecules. What's more, the 80-carbon buckyball itself was unusual: no one had ever before made one, either hollow or filled. Even though the metal-nitride molecules and the 80-carbon buckyball do not exist for long on their own, they stabilize each other in the new arrangement.

The buckyball has a net negative charge, while the metal cluster has a net positive one. This charge distribution of the metallic fullerene molecule gives it interesting properties, which researchers are still trying to understand.

The materials could be utilized as a more effective contrast agent in MRI, which is used to image soft body tissue such as the brain and spinal cord. Physicians currently inject gadolinium into a patient's body right before an MRI exam. The metal improves the resolution of the scans and increases the image contrast. But gadolinium is toxic, so it is wrapped with an organic compound. This does not eliminate the toxicity risk completely.

In contrast, the 80-carbon buckyball is a much stronger cage for trapping gadolinium nitride

Trapping other metals in the buckyball could lead to different applications like deveoping highly efficient solar cells.

Position Carbon Nanotube

Nantero (company) announced it has developed a method for positioning carbon nanotubes reliably on a large scale by treating them as a fabric which can be deposited using methods such as spincoating, and then patterned using lithography and etching.Nantero has also developed a method for purifying carbon nanotubes to the standards required for use in a production semiconductor fab, which means consistently containing less than 25 parts per billion of any metal contamination.

With these innovations, Nantero has become the first company in the world to introduce and use carbon nanotubes in mass production semiconductor fabs.

The company is developing NRAM—a high-density nonvolatile random access memory device intended for use as a universal memory.

Ultraviolet Light Reveals Secrets of Nanoscale Electronic Materials

An international team of scientists has used a novel technique to measure, for the first time, the precise conditions at which certain ultrathin materials spontaneously become electrically polarized. The research provides the fundamental scientific basis for understanding this "ferroelectric" state in materials needed for next-generation "smart card" memory chips and other devices.

A film of barium titanate whose thickness is just 4-tenths of a nanometer—or 4-hundred-millionths of a centimeter—can retain its ferroelectric properties when it is layered in thin sandwiches with non-ferroelectric layers of strontium titanate.

The scientists found that they could manipulate ferroelectricity by imposing different kinds of electrical and mechanical boundary conditions. The electrical conditions include the degree of resistance to polarization of the nonferroelectric material. The mechanical conditions included sandwiching ferroelectric layers between different layers of other materials, which mechanically restricts the movement of the atoms. By varying the thickness and composition of the nanoscale thin films, the researchers were able to change the phase-transition temperature by almost 500 Kelvin, obtaining ferroelectric properties more than 350 Kelvin—over 600 degrees Fahrenheit—above room temperature.

Nano Wires Detector

New Method Creates Nanowire Detectors Exactly Where Needed a team of investigators at the Nanosystems Biology Cancer Center has developed a method for creating conducting polymer nanowires in place within microfluidic circuitsThe researchers create the nanowires using standard microelectrodes built into the microfluidics device specifically for the purpose of carrying out electrochemical reactions within the channels of the device. This allows them to use the microfluidic channels to introduce the precursor molecules, or monomers, needed to create the conducting polymer nanowires and trigger an electrochemical reaction at the exact place where the nanowires are needed to function as biomolecule detectors. This reaction causes the monomers to link to one another, forming the conducting polymer nanowires. This process can create two different types of polymer nanowires, one made of polyaniline, the other of polypyrrole. The chemical reactions are completed within 40 minutes.

Once formed, the nanowires can function immediately as detectors, with the electrodes used to form the nanowires now functioning as the circuitry that connects the nanowires to electrical signal recorders. The investigators demonstrate that these detectors are highly sensitive to changes in pH and to changing ammonia concentrations, though they note that these nanowires should be able to be used to detect a wide range of biomolecules.

New detection systems using nanowires and microfluidics hold the promise of providing a quantum leap in the detection of cancer-related molecules and genes.

The work was done by researchers at the University of California, Los Angeles (UCLA), who claim that the result could find application in the development of bio-compatible electronics.It's an electronic device, fabricated from the tobacco mosaic virus conjugated with nanoparticles, which exhibits a unique memory effect, can be operated as an electrically bistable memory device whose conductance states can be controlled by a bias voltage. The states are non-volatile and can be digitally recognized.The TMV is a 300 nm tube consisting of a protein capsid (outer shell) and RNA core. According to the researchers, the TMV's thin, wire-like structure makes it suitable for attaching nanoparticles. In this case, it allowed them to add an average of sixteen positive platinum ions per virion. The device works by transferring charge, under a high electric field, from the RNA to the platinum nanoparticles with the TMV's surface proteins acting as an energy barrier, stabilising the trapped charges.

The TMV hybrid has an access time (the delay between a call for storing data and for data storing to begin) in the microsecond regime. This is comparable to today's flash memory. In addition, the device is non-volatile, which means that data is retained once the computer's power is turned off.

In the long term, these devices could one day be integrated in biological tissues for applications in therapeutics or biocompatible electronics.

Bistable Nanoswitch

The device is made of a free suspended multiwalled carbon nanotube interacting electrostatically with an underlying electrode. In the device circuit, there is a resistor in series with the nanotube, which plays an important role in the functioning of the device by adjusting the voltage drop between the nanotube and the underlying electrode.

A major advantage of the device is its geometry, which is fully compatible with current manufacturing techniques for mass production. The potential applications of the device include nano electromechanical systems (NEMS) switches, random-access memory elements and logic devices

Smart Paper

Nanocoating wood fibers results in smart paper layer-by-layer (LbL) assembly, is of great interest of its usage in the field of nanocoating. It allows creating nanometer-sized ultrathin films both on large surfaces and on microfibers and cores with the desired composition. Researchers at Louisiana Tech University have developed a simple and cost effective technique to fabricate an electrically conductive paper by applying layer-by-layer nanoassembly coating directly on wood microfibers during paper making process. Nanocoated wood microfibers and paper may be applied to make electronic devices, such as capacitors, inductors, and transistors fabricated on cost-effective lignocellulose pulp. The use of a conductive nanocoating on wood fibers can open the door for the future development of smart paper technology, applied as sensors, communication devices, electromagnetic shields, and paper-based displays.

New Method of Chemical Vapor Deposition for Smaller Nano Structures

Engineers at the California Institute of Technology have invented an ingenious new method for depositing tiny amounts of materials on surfaces. The researchers say that the technique, known as plasmon-assisted chemical vapor deposition, will add a powerful new tool to the existing battery of techniques used to construct microdevices.

New vapor deposition process can be used with a variety of materials by focusing a low-powered laser beam onto a substrate coated with gold nanoparticles. The laser wavelength is chosen to match a natural resonance in the gold particles, and those particles in the small spot illuminated by the laser (about one micron in diameter, or less than a hundredth the diameter of a human hair) absorb energy from the laser and quickly heat up, rising in temperature several hundred degrees.

The gold particles become hot enough to decompose precursor molecules in the gas that strike them, forming microscopic deposits on the nanoparticles. Since this only happens for the hot gold particles in the laser spot, and not for the nearby cool ones outside the laser spot, structures form only where the laser shines, allowing deposition in patterns "drawn" by moving the laser spot on the substrate.

The key to the process is the surprisingly low thermal conductivity at the tiny scales involved. The nanoparticles absorb energy from the laser very efficiently, but are much worse than bigger particles would be at getting rid of this energy by conducting heat away to the surroundings. As a result, the nanoparticles can be heated to temperatures much higher than expected based on classical concepts of heat conduction.

The ability to write micron-scale or smaller structures directly, without need for lithographic patterning and etching, while also keeping the substrate cool outside the small laser spot, opens up new possibilities for the types of structures that may be easily fabricated,

Breakthrough in Nanotube Manufacturing

Idaho Space Materials, Inc. has developed a revolutionary nanotube manufacturing process which produces uncontaminated, high purity single-walled carbon nanotubes (SWCNTs) at a very high production rate. While the high production rates achieved are a significant advancement, generating the material without harmful metal contaminants allows these nanotubes to be easily incorporated into the ever increasing applications for which this extraordinary material is used.Carbon nanotubes have drawn significant attention from the scientific community since their discovery in 1991 due to their very high tensile strength, excellent electrical conductivity, light weight, and outstanding thermal properties. Products enhanced by carbon nanotubes are reaching the consumer market in a wide variety of applications. Researchers in fields ranging from medicine, microelectronics, and advanced materials are discovering new uses for carbon nanotubes on a daily basis and are finding that they can dramatically improve our quality of life with this amazing material

Reducing the Size of Multi-Layer Nanoshells Enables New Sensing Applications

Nanoshells are a novel class of optically tunable nanoparticles that consist of alternating dielectric and metal layers. They have been shown to have tunable absorption frequencies that are dependent on the ratio of their inner and outer radii. Therefore nanoshells can potentially be used as contrast agents for multi-label molecular imaging, provided that the shell thicknesses are tuned to specific ratios. When used as contrast agents, nanoshells of small dimensions offer advantages in terms of delivery to target sites in living tissues, bioconjugation, steric hindrance, and binding kinetics. Besides their improved tissue penetration, smaller nanoshells generate a strong surface plasmon resonance and may exhibit absorption peaks in the visible—near-infrared spectrum. Sub-100 nm nanoshells also provide large surface areas to volume ratios for chemical functionalization that can be used to link multiple diagnostic (e.g. radioisotopic or magnetic) and therapeutic (e.g. anticancer) agents. Researchers at Northwestern University have come up with a relatively easy way to synthesize sub-100 nm nanoparticles that give rise to tunable peaks.

Recent Mie theory simulations suggest that small multi-layer gold nanoshells with diameters of around tens of nanometers can potentially generate a stronger surface plasmon resonance and exhibit ultra-sharp absorption peaks with spectrum widths of the order of 10 nm in the visible—near infrared spectrum, provided that the shells thicknesses are tuned to a certain ratio.

Atomically Modified Rice

A nanotech research initiative in Thailand aims to atomically modify the characteristics of local rice varieties—including the country's famous jasmine rice—and to circumvent the controversy over Genetically Modified Organisms (GMOs). Nanobiotech takes agriculture from the battleground of GMOs to the brave new world of Atomically Modified Organisms (AMOs).

The research involves drilling a nano-sized hole through the wall and membrane of a rice cell in order to insert a nitrogen atom. The hole is drilled using a particle beam (a stream of fast-moving particles, not unlike a lightening bolt) and the nitrogen atom is shot through the hole to stimulate rearrangement of the rice's DNA.

Nanofibre

It looks like simple piece of paper, but this small but powerful membrane is the prototype to a lighter gas mask used by healthcare or military workers in the future. Created by a group of National University of Singapore (NUS) researchers Made from nanofibres—fibres which are a hundred times smaller than a strand of human hair—the material is not only light, it can also neutralise chemical agents into non-toxic ones.

The $200,000 two-year project, which is funded by t he Defence Science and Technology Agency, also paves the way for lighter and more comfortable protective suits and gas masks used during chemical attacks. Traditional gear is heavy, uncomfortable, non-porous and need to be specially disposed of. There are also plans to use it in water and air purification treatment.

Nano Scale Thermometer

A nanoscale thermometer has recently been reported by Japanese researchers.

They made a nanothermometer by filling a carbon nanotube with liquid gallium. The new device works in air, unlike previous models, which only operated in vacuum. The thermometer, which is less than 150 nanometres in diameter, could find use in a range of micro-environmental applications.

Nanorobotic

"Nanorobotic" Arm to Operate Within DNA Sequence have developed a DNA cassette through which a nanomechanical device can be inserted and function within a DNA array, allowing for the motion of a nanorobotic arm first time scientists have been able to employ a functional nanotechnology device within a DNA array.

It is crucial for nanorobotics to be able to insert controllable devices into a particular site within an array, thereby leading to a diversity of structural states The invention has the potential to develop new synthetic fibres, advance the encryption of information and improve DNA-based computation.

Nanoparticle Drug Delivery Method to Replace Eye Drops

Eye drops could be a thing of the past thanks to an innovative and new method of delivering drugs to the eye using biodegradable materials. This new method, developed by a team of researchers led by biomaterials and drug delivery expert Dr John Tsibouklis at the University of Portsmouth, uses biodegradable polymer nanoparticles to administer drugs to the eye.

Biodegradable polymers can be combined with drugs in such a way that the drug is released into the eye in a very careful and controlled manner.

The drug would have to be placed into the eye just once.

The drug's release can be timed so it is constant, cyclic or triggered by an environmental or chemical signal, and the drug delivering polymer can be broken down naturally by the body when it is no longer needed.

People with eye conditions who use eye drops regularly would benefit from the biodegradable polymer drug delivery method. Eye drops have many disadvantages—two main ones being the need to administer drops regularly and low ocular bioavailability (too little of the drug is getting to areas of the eye most in need).

The common alternative option to eye drops, ophthalmic inserts, achieve sustained drug delivery but suffer from limitations also—they are difficult to insert, easy to misapply and are expensive to manufacture.

Nanotube Coating Promises Ice-Free Windscreens

A transparent lacquer containing carbon nanotubes could clear car windscreens or mirrors by acting as a heater. Thicker, opaque versions of the coating could turn entire floors of buildings into radiators, researchers claim.

The lacquer can be sprayed onto any surface and consists of a liquid base containing a mixture of nanotubes that conduct electricity. As the liquid dries, the nanotubes form a conducting network inside the lacquer. Passing a current through this network causes the layer to heat up.

In tests, a coating connected to a 12 volt power supply similar to car battery was able to clear ice from a plastic sheet in about 2 minutes although the test sheet was only the size of a paper back book.

The technology will replace the built-in wire filament heaters in car windscreens. As well providing more uniform heat than a filament heater, the nanotube film is more resistant to damage the coating could also be used for under-floor heating. A square metre of film about 0.3 millimetres thick provides around 15 kilowatts of heat—enough to heat a good sized hall.

8

IMPACT OF NANOTECHNOLOGY

SOCIETAL IMPACTS OF NANOTECHNOLOGY

Nanotechnology has a huge number and variety of applications across many different sectors. Potentially it could lead to more efficient and sustainable use of resources and have a beneficial impact for the vast majority of people throughout the world. However, as with all technologies there are also potential negative impacts on society. The main issues include privacy, social divide, communication, and risk.

Privacy

Ambient sensor systems can provide useful information such as pollution levels, traffic conditions and transmit this rapidly to portable devices, but they can also transmit information about individuals' activities. As such the potential for abuse is present and the limits on the type of information that can be captured and collated need to be clearly defined by society in general through the legislative system. Privacy issues may also arise through advances in medical diagnostics allowing doctors to routinely screen people for the presence of genetic disease. Should such technology be made compulsory to allow earlier treatment? If so, then what about a patient's right to choose? If not, then will health insurance companies demand it as a prerequisite? Who will have access to all this information and how will it be kept protected?

Social Divide

As with previous technologies such as IT, nanotechnology could have the effect of widening the divide between the rich and the poor, or more specifically the developed and developing world. Primarily this can be through advances in healthcare, transport, energy supplies, etc which may be more available to the wealthy. However, paradoxically it may also come about through a decreased use of natural resources. Many of the precious metals and minerals that new nanomaterials are expected to replace, and thereby reduce our dependency on non-renewables, are mined in the developing world. The loss of this revenue without a strategy for its replacement, will have a negative impact on the economy and development of these countries. To address these potential impacts, nanotechnology strategies are being exploited in different countries that specifically address that country's needs.

Communication

Acceptance of new developments and in particular the wide-ranging effects of nanotechnology, can only be achieved through communication and dialogue between scientists, industrialists, goverments and wider society. All too often in the past this has been ignored and resulted in misinformation and misunderstanding of the risks and benefits associated with the new development. This has been recognised by governments, research funding agencies and industry, and many now have initiatives that actively explore dialogue with social scientists and interested citizens, allowing the implications of new developments to be explained, concerns explored and opinions of different members of society incorporated into future planning.

Risk

Nanomaterials are being developed because they offer advantages to conventional materials. However, we still know little of the different effects they may have on human health and the environment compared with conventional materials. Several initiatives have been established by non-governmental organizations to consolidate existing published data, however to date there has been little concerted effort by governments. This is starting to change as independent experts advising government policy recommend that fundamental research into these potential risks be increased

No Spots

The clothing industry uses nanotech to make stain-repellent fabrics.

A chemical process during manufacture forces liquids to bead up when spilled on a garment for easy wiping away.

Socks that are made with nano-silver particles give anti-microbial protection, preventing bacteria and fungus that cause itchiness and smells.

SOCIAL AND ETHICAL IMPACTS

If it is difficult to predict the future direction of nanoscience and nanotechnologies and the timescale over which particular developments will occur, it is even harder to predict what will trigger social and ethical concerns. In the short to medium term concerns are expected to focus on two basic questions: 'Who controls uses of nanotechnologies?' and 'Who benefits from uses of nanotechnologies?' These questions are not unique to nanotechnologies but past experience with other technologies demonstrates that they will need to be addressed.

The perceived opportunities and threats of nanotechnologies often stem from the same characteristics. For example, the convergence of nanotechnologies with information technology, linking complex networks of remote sensing devices with significant computational power, could be used to achieve greater personal safety, security and individualised healthcare and to allow businesses to track and monitor their products. It could equally be used for covert surveillance, or for the collection and distribution of information without adequate consent. As new forms of surveillance and sensing are developed, further research and expert legal analysis might be necessary to establish whether current regulatory frameworks and institutions provide appropriate safeguards to individuals and groups in society. In the military context, too, nanotechnologies hold potential for both defence and offence and will therefore raise a number of social and ethical issues.

There is speculation that a possible future convergence of nanotechnologies with biotechnology, information and cognitive sciences could be used for radical human enhancement. If these possibilities were ever realised they would raise profound ethical questions.

HEALTH AND ENVIRONMENTAL IMPACTS

Health and Sanitation

Nanotechnology is already useful as a tool in health care research. In January 2005, researchers at the US Massachusetts Institute of Technology used 'optical tweezers'—pairs of tiny glass beads are brought together or moved apart using laser beams—to study the elasticity of red blood cells that are infected with the malaria parasite. The technique is helping researchers to better understand how malaria spreads through the body.

But nanotechnology could also one day lead to cheaper, more reliable systems for drug-delivery. For example, materials that are built on the nanoscale can provide encapsulation systems that protect and secrete the enclosed drugs in a slow and controlled manner. This could be a valuable solution in countries that don't have adequate storage facilities and distribution networks, and for patients on complex drug regimens who cannot afford the time or money to travel long distances for a medical visit.

Filters that are structured on the nanoscale offer the promise of better water purification systems that are cheap to manufacture, long-lasting and can be cleaned. Other similar technologies could absorb or neutralise toxic materials, such as arsenic, that poison the water table in many countries including India and Bangladesh.

Food Security

Tiny sensors offer the possibility of monitoring pathogens on crops and livestock as well as measuring crop productivity. In addition, nanoparticles could increase the efficiency of fertilisers. However, the Swiss insurance company SwissRe warned in a report in 2004 that they could also increase the ability of potentially toxic substances, such as fertilisers, to penetrate deep layers of the soil and travel over greater distances.

In addition, researchers in both developed and developing countries are developing crops that are able to grow under 'hostile' conditions, such as fields where the soil contains high levels of salt (sometimes due to climate change and rising sea levels) or low levels of water. They are doing this by manipulating the crops' genetic material, working on a nanotechnology scale with biological molecules.

Solar Panels

The application of nanotechnology in the field of renewable and sustainable energy (such as solar and fuel cells) could provide cleaner and cheaper sources of energy. These would improve both human and environmental health.

Tiny wastewater filters, for example, could sift emissions from industrial plants, eliminating even the smallest residues before they are released into the environment. Similar filters could clean up emissions from industrial combustion plants. And nanoparticles could be used to clean up oil spills, separating the oil from sand, removing it from rocks and from the feathers of birds caught in a spill.

Negative Impacts

Concerns have been expressed that the very properties of nanoscale particles being exploited in certain applications (such as high surface reactivity and the ability to cross cell membranes) might also have negative health and environmental impacts. Many nanotechnologies pose no new risks to health and almost all the concerns relate to the potential impacts of deliberately manufactured nanoparticles and nanotubes that are free rather than fixed to or within a material. Only a few chemicals are being manufactured in nanoparticulate form on an industrial scale and exposure to free manufactured nanoparticles and nanotubes is currently limited to some workplaces (including academic research laboratories) and a small number of cosmetic uses. We expect the likelihood of nanoparticles or nanotubes being released from products in which they have been fixed or embedded (such as composites) to be low but have recommended that manufacturers assess this potential exposure risk for the lifecycle of the product and make their findings available to the relevant regulatory bodies.

Few studies have been published on the effects of inhaling free manufactured nanoparticles and we have had to rely mainly on analogies with results from studies on exposure to other small particles– such as the pollutant nanoparticles known to be present in large numbers in urban air, and the mineral dusts in some workplaces. The evidence suggests that at least some manufactured nanoparticles will be more toxic per unit of mass than larger particles of the same chemical. This toxicity is related to the surface area of nanoparticles (which is greater for a given mass than that of larger particles) and the chemical reactivity of the surface (which could be increased or decreased by the use of surface coatings). It also seems likely that nanoparticles will penetrate cells more readily than larger particles.

It is very unlikely that new manufactured nanoparticles could be introduced into humans in doses sufficient to cause the health effects that have been associated with the nanoparticles in polluted air. However, some may be inhaled in certain workplaces in significant amounts and steps should be taken to minimise exposure. Toxicological studies have investigated nanoparticles of low solubility and low surface activity. Newer nanoparticles with characteristics that differ substantially from these should be treated with particular caution. The physical characteristics of carbon and other nanotubes mean that they may have toxic properties similar to those of asbestos fibres, although preliminary studies suggest that they may not readily escape into the air as individual fibres. Until further toxicological studies have been undertaken, human exposure to airborne nanotubes in laboratories and workplaces should be restricted.

If nanoparticles penetrate the skin they might facilitate the production of reactive molecules that could lead to cell damage. There is some evidence to show that nanoparticles of titanium dioxide (used in some sun protection products) do not penetrate the skin but it is not clear whether the same conclusion holds for individuals whose skin has been damaged by sun or by common diseases such as eczema. There is insufficient information about whether other nanoparticles used in cosmetics (such as zinc oxide) penetrate the skin and there is a need for more research into this. Much of the information relating to the safety of these ingredients has been carried out by industry and is not published in the open scientific literature.

There is virtually no information available about the effect of nanoparticles on species other than humans or about how they behave in the air, water or soil, or about their ability to accumulate in food chains. Until more is known about their environmental impact we are keen that the release of nanoparticles and nanotubes to the environment is avoided as far as possible. As a precautionary measure factories and research laboratories treat manufactured nanoparticles and nanotubes as if they were hazardous and reduce them from waste streams and that the use of free nanoparticles in environmental applications such as remediation of groundwater be prohibited.

There is some evidence to suggest that combustible nanoparticles might cause an increased risk of explosion because of their increased surface area and potential for enhanced reaction. Until this hazard has been properly evaluated this risk should be managed by taking steps to avoid large quantities of these nanoparticles becoming airborne.

Nanoparticles found in pollution from traffic exhaust but also used in making household goods such as paint, sunblock, food, cosmetics and clothes can cause damage to the cells of the liver experiments have shown that nanoparticles delivered into the lungs crossed the lung barrier and entered the blood. Particles in the blood can reach the liver, and it is also known that nanoparticles directly injected into the blood for medical purposes are also likely to end up in the liver.

Researchers don't know if the nanoparticles are safely eliminated from the liver by specialised cells or whether these extremely small particles can enter the liver cells and disrupt their normal functioning.

Research into the hazards and exposure pathways of nanoparticles and nanotubes is required to reduce the many uncertainties related to their potential impacts on health, safety and the environment. This research must keep pace with the future development of nanomaterials.

RISKS WITH NANOTECHNOLOGY

- Economic disruption from an abundance of cheap products
- Economic oppression from artificially inflated prices
- Personal risk from criminal or terrorist use
- Personal or social risk from abusive restrictions
- Social disruption from new products/lifestyles

- Unstable arms race
- Collective environmental damage from unregulated products
- Free-range self-replicators (grey goo)
- Black market in nanotech (increases other risks)
- Competing nanotech programs (increases other risks)
- Attempted relinquishment (increases other risks)

Some of these risks arise from too little regulation, and others from too much regulation. Several different kinds of regulation will be necessary in several different fields. An extreme or knee-jerk response to any of these risks will create fertile ground for other risks. The temptation to impose apparently obvious and simple solutions to problems in isolation must be avoided.

Disruption of the basis of economy is a strong possibility. The purchaser of a manufactured product today is paying for its design, raw materials, the labor and capital of manufacturing, transportation, storage, and sales. Additional money—usually a fairly low percentage—goes to the owners of all these businesses. If personal nanofactories can produce a wide variety of products when and where they are wanted, most of this effort will become unnecessary. This raises several questions about the nature of a post-nanotech economy. Will products become cheaper ? Will capitalism disappear ? Will most people retire or be unemployed ? The flexibility of nanofactory manufacturing, and the radical improvement of its products, imply that non-nanotech products will not be able to compete in many areas. If nanofactory technology is exclusively owned or controlled, will this create the world's biggest monopoly, with extreme potential for abusive anti-competitive practices? If it is not controlled, will the availability of cheap copies mean that even the designers and brand marketers don't get paid ? Much further study is required, but it seems clear that molecular manufacturing could severely disrupt the present economic structure, greatly reducing the value of many material and human resources, including much of our current infrastructure. Despite utopian post-capitalist hopes, it is unclear whether a workable replacement system could appear in time to prevent the human consequences of massive job displacement.

Major investment firms are conscious of potential economic impact. In the mainstream financial community, there is growing recognition that nanotechnology represents a significant wave of innovation with the potential to restructure the economy. Nano-built products may be vastly overpriced relative to their cost, perpetuating unnecessary poverty. By today's commercial standards, products built by nanofactories would be immensely valuable. A monopoly would allow the owners of the technology to charge high rates for all products, and make high profits. However, if carried to its logical conclusion, such a practice would deny cheap lifesaving technologies (as simple as water filters or mosquito netting) to millions of people in desperate need. Competition will eventually drive prices down, but an early monopoly is likely for several reasons. Due to other risks listed on this page, it is unlikely that a completely unregulated commercial market will be allowed to exist. In any case, the high cost of development will limit the number of competing projects. Finally, a company that pulls ahead of the pack could use the resulting huge profits to stifle competition by means such as broad enforcement of expansive patents and lobbying for special-interest industry restrictions.

The price of a product usually falls somewhere between its value to the purchaser and its cost to the seller. Molecular manufacturing could result in products with a value orders of magnitude higher than their cost. It is likely that the price will be set closer to the value than to the cost; in this case, customers will be unable to gain most of the benefit of "the nanotech revolution". If pricing products by their value is accepted, the poorest people may continue to die of poverty, in a world where products costing literally a few cents would save a life. If (as seems likely) this situation is accepted more by the rich than by the poor, social unrest could add its problems to untold unnecessary human suffering. A recent example is the agreement the World Trade Organization was working on to provide affordable medicines to poor countries—which the Bush administration partially prevented (following heavy lobbying by American pharmaceutical companies) despite furious opposition from every other WTO member.

Criminals and terrorists could make effective use of the technology. Criminals and terrorists with stronger, more powerful, and much more compact devices could do serious damage to society. Defenses against these devices may not be installed immediately or comprehensively. Chemical and biological weapons could become much more deadly and easier to conceal. Many other types of terrifying devices are possible, including several varieties of remote assassination weapons that would be difficult to detect or avoid.

As a result of small integrated computers, even tiny weapons could be aimed at targets remote in time and space from the attacker. This will not only impair defense, but also will reduce post-attack detection and accountability. Reduced accountability could reduce civility and security, and increase the attractiveness of some forms of crime.

If nanofactory-built weapons were available from a black market or a home factory, it would be quite difficult to detect them before they were launched; a random search capable of spotting them would almost certainly be intrusive enough to violate current human rights standards.

Nanotech weapons would be extremely powerful and could lead to a dangerously unstable arms race. Molecular manufacturing raises the possibility of horrifically effective weapons. As an example, the smallest insect is about 200 microns; this creates a plausible size estimate for a nanotech-built antipersonnel weapon capable of seeking and injecting toxin into unprotected humans. The human lethal dose of botulism toxin is about 100 nanograms, or about 1/100 the volume of the weapon. As many as 50 billion toxin-carrying devices—theoretically enough to kill every human on earth—could be packed into a single suitcase. Guns of all sizes would be far more powerful, and their bullets could be self-guided. Aerospace hardware would be far lighter and higher performance; built with minimal or no metal, it would be much harder to spot on radar. Embedded computers would allow remote activation of any weapon, and more compact power handling would allow greatly improved robotics. These ideas barely scratch the surface of what's possible.

An important question is whether nanotech weapons would be stabilizing or destabilizing. Nuclear weapons, for example, perhaps can be credited with preventing major wars since their invention. However, nanotech weapons are not very similar to nuclear weapons. Nuclear stability stems from at least four factors. The most obvious is the massive destructiveness of all-out nuclear war. All-out nanotech war is probably equivalent in the short term, but nuclear weapons also have a high long-term cost of use (fallout, contamination) that would be much lower with nanotech weapons. Nuclear weapons cause indiscriminate destruction; nanotech weapons could be targeted. Nuclear weapons require massive research effort and industrial development, which can be tracked far more easily than nanotech weapons development; nanotech weapons can be developed much more rapidly due to faster, cheaper prototyping. Finally, nuclear weapons cannot easily be delivered in advance of being used; the opposite is true of nanotech. Greater uncertainty of the capabilities of the adversary, less response time to an attack, and better targeted destruction of an enemy's visible resources during an attack all make nanotech arms races less stable. Also, unless nanotech is tightly controlled, the number of nanotech nations in the world could be much higher than the number of nuclear nations, increasing the chance of a regional conflict blowing up. Military applications of molecular manufacturing have even greater potential than nuclear weapons to radically change the balance of power.

Molecular manufacturing will reduce economic influence and interdependence, encourage targeting of people as opposed to factories and weapons, and reduce the ability of a nation to monitor its potential enemies. It may also, by enabling many nations to be globally destructive, eliminate the ability of powerful nations to "police" the international arena. By making small groups self-sufficient, it can encourage the breakup of existing nations.

Grey goo was an early concern of nanotechnology. When nanotechnology-based manufacturing was first proposed, a concern arose that tiny manufacturing systems might run amok (run about in frenzied thirst) and 'eat' the biosphere, reducing it to copies of themselves. In 1986, Eric Drexler wrote, "We cannot afford certain kinds of accidents with replicating assemblers." More recent designs by Drexler and others make it clear, though, that replicating assemblers will not be used for manufacturing— nanofactories will be much more efficient at building products, and a nanofactory is nothing like a 'grey goo' robot.

Grey goo would entail five capabilities integrated into one small package. These capabilities are: Mobility—the ability to travel through the environment; Shell—a thin but effective barrier to keep out diverse chemicals and ultraviolet light; Control—a complete set of blueprints and the computers to interpret them (even working at the nanoscale, this will take significant space); Metabolism—breaking down random chemicals into simple feedstock; and Fabrication—turning feedstock into nanosystems. A nanofactory would use tiny fabricators, but these would be inert if removed or unplugged from the factory. The rest of the listed requirements would require substantial engineering and integration.

Grey goo won't happen by accident, but eventually could be developed on purpose. Although grey goo has essentially no military and no commercial value, and only limited terrorist value, it could be used as a tool for blackmail. Cleaning up a single grey goo outbreak would be quite expensive and might require severe physical disruption of the area of the outbreak (atmospheric and oceanic goos deserve special concern for this reason). Another possible source of grey goo release is irresponsible hobbyists. The challenge of creating and releasing a self-replicating entity apparently is irresistible to a certain personality type, as shown by the large number of computer viruses and worms in existence. We probably cannot tolerate a community of "script kiddies" releasing many modified versions of goo.

Development and use of molecular manufacturing poses absolutely no risk of creating grey goo by accident at any point. However, goo type systems do not appear to be ruled out by the laws of physics, and we cannot ignore the possibility that the five stated requirements could be combined deliberately at some point, in a device small enough that cleanup would be costly and difficult. Drexler's 1986 statement can therefore be updated: We cannot afford criminally irresponsible misuse of powerful technologies. Having lived with the threat of nuclear weapons for half a century, we already know that.

Facing all these risks, there will be a strong temptation simply to outlaw the technology. However, we don't believe this can work. Many nations are already spending millions on basic nanotechnology; within a decade, advanced nanotech will likely be within the reach of large corporations. It can't be outlawed worldwide. And if the most risk-aware countries stop working on it, then the less responsible countries are the ones that will be developing it and dealing with it. Besides, legal regulation may not have much effect on covert military programs.

Molecular manufacturing may be delayed by strict regulation, but this would probably make things worse in the long run. If MM development is delayed until it's relatively easy, it will then be a lot harder to keep track of all the development programs. Also, with a more advanced technology base, the development of nano-built products could happen even faster than we have described, leaving less time to adjust to the societal disruptions.

9
GLOBAL SCENARIO IN NANOTECHNOLOGY

Nanotechnology is expected to be one of the most important technologies of this century because if offers solutions to a variety of health and environmental problems. Moreover, new nanomaterials and nanodevices will have a major impact in many areas of the global economy. Indeed, analysts estimate that the market for nanotechnology products will increase to $2,600 billion by 2014, and that ten million new jobs will be created in areas of manufacturing related to nanotechnology by then.

Nanotechnology would exceed $1,000 billion by end 2010 in the world economy. The increase in demand for nanoscale materials, tools and devices would reach $28.7 billion in as soon as 2008. These would help medical scientists, engineers and other researchers to invent and innovate in the fields of health, IT, communications and consumer goods.

Companies like Intel, IBM, DuPont, 3M, General Electric, Samsung and Hitachi have spent $3.8 billion in 2004. This has been divided into 46 per cent by nanotechnology companies in North America, 36 per cent by Asian companies, 17 per cent by European companies and about 1 per cent by rest of the world.

Scientists, researchers, business managers, investors, funding agencies and governments worldwide all acknowledge the huge social and economic potential of nanotechnology, which is why public funding has increased from 500 million in 1999 to almost 4 billion in 2004. Scientific publications in nanotechnology have increased by a factor of six over the past ten years, and the number of nanotechnology patents has also increased substantially. However, the rate of growth has varied across the globe, and also between different areas of nanoscience and nanotechnology.

Scientific publications are the most appropriate indicator for measuring scientific activity or output. Data from the Science Citation Index, which is compiled by Thomson, show that researchers in the United States publish many more papers on nanoscience and nanotechnology than researchers from any other country. Indeed, during the period 1999-2004, US scientists published more than 18,000 nano papers, compared with just over 8,000 papers from Japan, and around 7,000 papers each from China and Germany. The UK is in fifth place, followed by France, Italy, South Korea, Canada and Spain. These rankings change when nanoscience is broken into different subfields. Within the subfield of nanomaterials, for instance, China leapfrogs Japan to take second place behind the US.

However, some scientific papers have a bigger impact than others (as measured by the number of times they are cited by other papers), and when nations are ranked by 'impact factor', two small European countries are in the lead: Switzerland is first, followed by the Netherlands and then the US. Canada moves up to fourth place, followed by Belgium, Ireland, the UK, Denmark, France and Japan. Germany comes a surprising eleventh place in this ranking.

Patents reflect the ability of companies and nations to transfer scientific results into technological applications and economic development. By analysing data from patent offices around the world, it is therefore possible to identify those areas of nanotechnology that are most likely to have a economic or commercial impact, and those individuals, organizations and nations that are going to benefit most.

The European Patent Office (EPO) has developed a methodology to identify and classify ('tag') nanotechnology patents and patent families. Nanoelectronics and nanomaterials were the two most active areas in terms of patents, accounting for more than half the total number of patent applications. The other four areas in the EPO classification scheme are nanomagnetics, nano-optics, nanobiotechnology and nanodevices.

The number of nanotechnology patents more than doubled between 1995 and 2003, but the growth has not been as spectacular as one might expect, with the peak of 1999 being followed by a fall in numbers, and a year of zero growth following the peak of 2002. Growth has been strongest in the Americas (mainly the US and

Canada), but there have been occasional year-on-year decreases. Growth has been slower but more consistent in Asia (mainly Japan and South Korea) and Europe (mainly Germany, the UK, France and the Netherlands)

It is obvious that the Americas is by far the most active region in the world for registering patents in nanotechnology, accounting for half the patent applications each year during the period covered, and also being responsible for the peaks in global patent applications witnessed in 1999 and 2002. In the same time period, Europe's share never exceeded 20 per cent. When the data is analysed by nation, rather than region, the US comes out top overall, and also in each of the six subfields, followed by Japan, Germany, the United Kingdom and France. Within nanobiotechnology, Germany, France and Canada are the strongest after the US, while the Netherlands and Sweden, India do well in nanoelectronics. Elsewhere, Belgium and Taiwan rank high in nanomaterials, Switzerland is strong in nanodevices, and the UK performs well in nano-optics.

The number of patent applications increased, on average, by 14 per cent per year between 1995 and 2003, with the growth being faster in the first half of this period. There were, however, huge differences between the six different subfields. Nanoelectronics, nanomaterials, nanodevices and nanomagnetics had the highest growth rates in the 1990s, but growth slowed down between 1999 and 2003—and even became negative for nanodevices. In contrast, nanobiotechnology and nano-optics both experienced negative growth in the late 1990s, but the rate increased to around 20 per cent per year after 2000. However, in absolute terms, both these areas remain much smaller than nanoelectronics and nanomaterials.

The Funding Base

The launch of the National Nanotechnology Initiative in the US in 1999 is generally seen as the start of the global race to exploit all the possibilities offered by nanotechnology. However, funding for nanoscience was already established in many regions of the world by this time, with Europe already being strong in nanomaterials by the mid-1980s.

As with patents and publications, the US leads the way in

public spending on nanotechnology, with the federal government investing 910 million in 2004, followed by Japan, the European Commission (EC), the individual states in the US, and Germany . Although China is ninth on this list, if purchasing power was taken into account, it would rank much higher. And when spending by the EC and the EU member states are added together, the total exceeds that of the federal and state governments in the US.

In the US the figure for the total spending on nanotechnology research rises to almost 3 billion when industry sources are included, followed by 2.3 billion for Japan and less than 2 billion for Europe . In other words, whereas industry sources account for around 60 per cent of total nanotechnology spending in the US, in Europe the corresponding figure is only one-third.

Nanotechnology in Asia Pacific

Since the announcement of the US National Nanotechnology Initiative in January 2000, the Science and Technology Policies of Asia Pacific countries have undergone significant changes.

Nanotechnology is now one of the high-priority areas for Asia Pacific governments as well. Budgets for the new technology research and development have been raised largely and more strategically allocated. Total expenditure for Asia Pacific countries has exceeded $1 billion for the past 2 years and will continue to increase.

Japan has been investing in nanoscience since the 1980s and after the USA is the heaviest nanotechnology R&D investor worldwide, and along with China, South Korea and Taiwan has increased its budget substantially since 2001.

Australia is currently evolving a national strategy, but already has considerable infrastructure and funding in place. India, Thailand and Vietnam have also identified nanotechnology as a significant area and are moving towards implementing the proper framework.

Frontier for Nanotechnolgy

Advanced Nano-tech Lab Opens

HCM CITY—Viet Nam's largest and most advanced nano

technology lab was inaugurated 23-12-06 at the HCM City National University.

The lab was built at a cost of US$4.5 million, according to one of the country's leading physicists, Prof. Nguyen Van Hieu. The lab will conduct research into nano-based technology and materials, provide research services and train scientists in the field.

The lab has a co-operation agreement with 13 universities, institutes and research centres in the US, France, the Netherlands, Switzerland, the Republic of Korea and Singapore

Berkeley to be First City to Regulate Nanotechnology

San Francisco: The use of subatomic materials as microscopic building blocks for thousands of consumer products has turned into a big business so quickly that few are monitoring the so-called nanotechnology's effects on health and the environment.

So Berkeley intends to be the first city to step into the breach and attempt to regulate the nascent but fast-growing industry.

The City Council is expected to amend its hazardous materials law to compel researchers and manufacturers to report what nanotechnology materials they are working with and how they are handling the tiny particles.

The aim of nanotechnology, in the commercial world, is to develop new products and materials by changing or creating materials at the atomic and molecular level. But much of the impacts from those developments remains unknown, particularly with regard to possible environmental and health problems.

STATISTICS

R&D Investment in Nanotechnology Worldwide

- Worldwide R&D expenditure on nanotechnology has increased from US $432 million in 1997 to US $4.081 billion in 2005.
- European Union: US $126 million in 1997; US $1.05 billion in 2005.
- US: $116 million in 1997; $1.081 billion in 2005, 2006 $1,301 million

- Japan: US $120 million in 1997; US $950 million in 2005.
- All others: US $70 million in 1997; US $1 billion in 2005.
- India: Since October 2001, when the Nano Science and Technology initiative was launched, India has invested Rs. 200 crore (about US $44.5 million). During 2006-07, Rs. 180 crore is the projected expenditure.

Planned Spending

- Japan $200,000 million in total by 2010
- US: $ 2,027 million each year (until 2008)
- Germany: $ 201 million in 2005
- China: $ 100 million each year
- UK: $ 45 million each year (2003-2009)

National Science Foundation Estimates

Global market for nanotechnology products in 2015 $1 trillion Global Workforce required in Nanotechnology Industries by 2018, 2 million.

INDIAN SCENARIO

Indian nanotechnology is estimated to be USD 100 million and is estimated to grow at over 35 per cent per year. The Government of India is planning for large investments in R&D through several initiatives and the private companies will exploit the technology for commercial benefits.

India is 20 years behind the U.S. and Europe in the field of nanotechnology because of lack of funding and short-sightedness on the part of the Government.

Unfortunately even the present situation for work in nanotechnology in India is poor because of the continued lack of funding, sub-optimal centres and lack of industry initiative in the areathe average global spending in the field over the past four years has been $13 billion, while India's spending has been almost negligible.

Some institutes like IIT Bombay, Banaras Hindu University, IISc Bangalore and few others are doing really a good job.

Let us take some examples on Products progress in India

(a) *Water: Nano Tube Filter—Water Purification*

The scientists from Banaras Hindu University have devised a simple method to produce carbon nanotube filters that efficiently remove micro-to nano-scale contaminants from water and heavy hydrocarbons from petroleum. Made entirely of carbon nanotubes, the filters are easily manufactured using a novel method for controlling the cylindrical geometry of the structure. The work was supported in part by the Ministry of Human Resource Development and Department of Science and Technology in India

The filters are hollow carbon cylinders several centimeters long and one or two centimeters wide with walls just one-third to one-half a millimeter thick. They are produced by spraying benzene into a tube-shaped quartz mold and heating the mold to 900°C. The nanotube composition makes the filters strong, reusable, and heat resistant, and they can be cleaned easily for reuse.

The carbon nanotube filters offer a level of precision suitable for different applications. They can remove 25-nanometer-sized polio viruses from water, as well as larger pathogens, such as E. coli and Staphylococcus aurous bacteria. The researchers believe this could make the filters adaptable to micro fluidics applications that separate chemicals in drug discovery.

This is a classic application of the latest in science—Nano science, to age old problem of water purification. If properly used, this can help in lessoning the burden in our drinking water missions leading to the availability of safe drinking water that will result in minimizing the water borne diseases.

(b) *Healthcare: Typhoid Detection Kit*

Typhoid Detection Kit has been developed by DRDE, Gwalior using the nano sensor developed by Prof. A.K. Sood, and his team from IISc, Bangalore. Typhoid fever caused by Salmonella typhi is a major health problem and an important challenge to health authorities of third world countries due to unsatisfactory water supply, poor sanitary conditions, malnutrition, emergence of antibiotic resistant strains etc. According to an estimate the worldwide incidence to typhoid fever is 16 million cases annually and death rate is 6 lakhs individual per year worldwide. In India, the morbidity due to typhoid varies from 102 to 2219/100,000 population in different parts of the countries. In some areas typhoid fever is responsible for 2-5 per cent of all deaths.

In India for routine diagnosis for typhoid disease Widal test is performed with single serum sample which does not provide the correct diagnosis of infection. Therefore a Latex agglutination based test has been developed at DRDE, Gwalior using recombinant DNA technology and immunological technique for rapid diagnosis of typhoid infection. The test detects S. typhi antigen directly in patient's serum within 1-3 minutes which is very important for initiating early treatment and saving human life.

A collaborative work has been carried out with Prof. A.K. Sood of Indian Institute of Science, Bangalore, the sensitivity of the test has been increased 30 times by applying a small electric charge (1.5 V). With this improvement, extreme low concentrations of the antigen in clinical sample can be detected. Moreover, very small quantity of clinical sample as low as 2-3 µl is required to perform the above test as compared to 10-15 µl sample required for latex agglutination test.

(c) *Power: Gas Flow Induced Generation of Voltage from Solids*

Prof A.K. Sood, professor of Physics at IISc and his student Shankar Ghosh has studied, experimented and found that the liquid flow in carbon nano tubes can generate electric current. One of the most exciting applications to emerge from the discovery is the possibility of a heart pacemaker—like device with nanotubes, which will sit in the human body and generate power from blood. Instead of batteries, the device will generate power by itself to regulate defective heart rhythm. The IISc has transferred the exclusive rights of the technology to an American start-up Trident Metrologies. They will develop the prototypes and commercialize the gas flow sensors.

(d) *Drug Delivery System*

A research group headed by Professor A.N. Maitra of the University of Delhi's Chemistry Department has developed 11 patentable technologies for improved drug delivery systems using nanoparticles. Four of these processes have been granted U.S. patents. One of the important achievements at the initial stage of drug delivery research was development of a reverse micelles based process for the synthesis of hydrogel and 'smart' hydrogel nanoparticles for encapsulating water-soluble drugs. This method enabled one to synthesize hydrogel nanoparticles of size less than 100nm diameter. This technology has been sold to Dabur Research Foundation in 1999.

Another technology has been transferred to industry deals with nanoparticle drug delivery for eye diseases. Traditionally, steroids have been used extensively in the treatment of ocular inflammatory disease and allergies. However, prolonged use of steroids has many side effects. The Delhi University group's process uses nanoparticles to encapsulate non-steroidal drugs. "This process improves the bioavailability of the drug on the surface of the cornea". The technology has been transferred to Chandigarh-based Panacea Biotech Ltd.

(e) *Microwave CNTs Production Unit*

DMSRDE, Kanpur is synthesizing non-aligned, quasi-aligned and aligned CNT with a batch size of 50 grams using a fast synthesis process. It has a maximum operating temperature 12000°C. The CNTs will have applications in EM absorbers, composites, gas sensors, flow monitors, field emission devices

What India can do According to Honarable President of India Dr. A.P.J. Abdul Kalam

There have been some excellent areas of basic research in very advanced topic right from late 1950s and 60s. For example we had excellent work on aero science, material science and also excellent work on semi-conductor research. Some world class scientists published papers also. Later even in the Liquid Crystal Display (LCD) area some original work was done by Raman Research Institute. We should now work to realize world class materials.

However, when one studies the technological and commercial aspects of these areas, the results have not been commensurate. We never produced an indigenous aircraft until very recently.

Also we missed micro electronic revolution in the 70s. Similarly, in Liquid Crystal Display, we never even started a good commercial production whereas countries like Taiwan and South Korea have become world leaders in commercial production of these products. If we study, many areas, such trend seems to be the pattern.

We should not repeat this again in areas like Nano-Technology and Convergence of nano-technology with ICT and BT. Therefore, even while we are concentrating on basic research with eminent scientists working in it, simultaneously Indian industrial group small, big and medium should concurrently work on commercialization of nano-technologies. It may well be that the technologies are developed in India or in USA or in other countries. The main focus should be speedy commercialization to fit into the global market. The time is now ripe since our economy is in the ascent phase and the manufacturing sector has established adequate capacity to promote rapid commercialization of products. Towards this, I suggest that separate funds which can be used primarily for commercialization by Indian industry either for in-house or partnership mode for specific product should be encouraged at the earliest. And innovative managerial mechanism has to be there to utilize these funds with utmost speed and with commercial success in mind. The project should be market driven. I would suggest to the government to set apart sufficient funds on non-lapsable basis for nano-technology commercialization, and I would like industries to take a lead in this matter. Simultaneously, there is a need to take action to build the human resource required for undertaking the challenging tasks in this new sector. I would recommend the nano-technology community, particularly, the industrialists to read the book titled "Nano-Technology Market and Company Report"—finding hidden pearls prepared by Deutsche Bank and other experts, with details of over 350 nanocompanies and interviews with over 100 entrepreneurs.

We should mount a mission mode operation to deliver tangible products to meet our national demand as well as to be beneficial to the other countries. Some of the national missions and the possible areas of research, design, development and production of products with reference to the areas of importance for development using nano science and technologies. (Source : HT)

Agriculture and Food Processing

We are in the mission of generating 400 tones of food grains with reduced land, with reduced water and with reduced workforce. It is essential to take an agro food processing in a big way which will bring employment potential in rural areas.

Some of the possible areas of research in agriculture and food processing are: Nano-porous zeolites for slow-release and efficient doses of water and fertilizers for plant, and of nutrients and drugs for livestock, nano-capsules for herbicide delivery, nano sensors for soil quality and for plant health monitoring. Nano-composites for plastic film coatings used in food packaging, antimicrobial nano-emulsions for applications in decontamination of food equipment, packaging or food processing are other important areas of research.

Infrastructure

India is aspiring to build hundred million homes within next ten years. The infrastructure development in metropolitan and tier-2 cities needs to be enhanced in the form of new bridges, airports, marketing complexes and industrial units. 40 per cent of the rural areas need to be covered with all weather roads; we need to double the present national highways ratio for the 100 square kilometer area.

Can nano-science material and technology research provide a solution? Our research focus in the nano-material should be towards cheaper rural housing, surfaces, coatings, use of concrete with heat and light exclusion; Can we develop heat resistance nano-material to block ultraviolet and infra red radiation; Can we develop a nano-molecular structure to make concrete more robust to water seepage, Can we have self cleaning surfaces with bio active coating?

Energy

Energy Independence has to be our nation's first and highest priority. We must be determined to achieve this within the next 25 years i.e by the year 2030. When our population may touch 1.4 billion people, demand from power sector will increase from the existing 120,000 MW to about 400,000 MW. This assumes an energy growth rate of 5 per cent per annum. Electric power generation in India now accesses four basic energy sources: Fossil fuels such as oil, natural gas and coal; Hydroelectricity; Nuclear power; and Renewable energy sources such as bio-fuels, solar, biomass, wind and ocean.

Fortunately for us, 89 per cent of energy used for power generation today is indigenous, from coal (56%), hydroelectricity (25%), nuclear power (3%) and Renewable (5%). Solar energy segment contributes just 0.2 per cent of our energy production. Thus it would be seen that only 11 per cent of electric power generation is dependent on oil and natural gas which is mostly imported at enormous cost. Only 1 per cent of oil is (about 2-3 million tonnes of oil) being used every year for producing electricity. However, power generation to the extent of 10 per cent is dependent on high cost gas supplies. The most significant aspect, however would be that the power generated through renewable energy technologies may target 20 to 25 per cent against the present 5 per cent.

Keeping this requirement in mind, can we find an innovative solution for the use of nano-technology using higher efficiency CNT based solar photovoltaic cells with an efficiency of 45 per cent in partnership with Penn State University? This will enable setting up of modular hundred megawatts solar SPV plants across the country in a reduced land with reduced cost when compared to present figure of Rs. 20 Crore per megawatt SPV plant with 14 per cent efficiency photovoltaic cells. Can we also develop novel hydrogen storage systems based on carbon nano tube for energy storage?

Safe Drinking Water

Over 50 per cent people do not have safe drinking water in India. Shall we embark on a mission for water purification, water de-toxification, water desalination through nano membranes and nano sensor for detecting contaminants and pathogens? How the nano-porous zeo-lites, nano-porous polymers can be used to design and develop products for water purification.

Healthcare

India has already patented the development of drug delivery

system using nano-technology. Stem cell research in India is advancing in the field of cardiology, ophthalmology, diabetic research, endocrinology, oncology and immunology. It is essential to develop drug delivery system for stem cell implantation into the specific organs of the body related to the ailment using nano technology.

Aero Space

Emerging technologies such as MEMS, Nano, Information technology, biotechnology, space research, Hypersonic, High power lasers and microwave will be dominating the future in every field and applications. The advancements in material science and technology will give a major thrust to the realization of advanced aerospace systems. We are today at the convergence of Nano, Bio and Information technologies, that will lead to new generation aero space devices and products.

Molecular nano technology has enormous potential for future aerospace systems. Research has shown that newly discovered class of molecules, particularly carbon nano tubes built from graphite sheets curved into a wide variety of close shapes, may lead to tougher, high temperature materials that can survive in vacuum and other harsh environments. Carbon nano tubes are normal form of carbon with remarkable electrical and mechanical properties. It is hoped that such materials could revolutionize electronic design and open the space frontier by radically lowering the cost of launch to orbit.

Carbon Nano Tubes reinforced with polymer matrix will result in compo-sites which are super strong, light weight, small and intelligent structures in the field of material science. This has tremendous aerospace applications.

ICT

Molecular switches and circuits along with nano cell will pave the way for the next generation computers. Ultra dense computer memory coupled with excellent electrical performance will result in low power, low cost, nano size and yet faster assemblies. This will result in the small scale assembly of computers, tablet PCs, display systems etc.,

With the emergence of Nanotechnology, there is convergence of nano-bio-info technologies resulting new devices which has wider applications in structure, electronics, and healthcare and space systems. Potential applications are virtually endless. Progress in nanotechnology is spurred by collaboration among researchers in material science, mechanical engineering, computer science, molecular biology, physics, electrical engineering, chemistry, medicine and aerospace engineering. This is one of the important emerging area which brings synergy in research and development by combining the strengths of the multiple domain knowledge leading to the creation of knowledge society. Our educational institutions and universities should have a special purpose missions based on their core competence.

10

FUTURE OF NANOTECHNOLOGY

What could the Future Hold for Nanotechnology ?

With so many unknowns it is diffi cult to have a meaningful discussion about the future of nanotechnology. The following scenarios are plausible, internally consistent, possible futures, which can be used to explore possible developments. They are explicitly not predictive, but should be used as qualitative planning and communication tools. A scenario-building exercise is not intended to create good versus bad scenarios or likely versus unlikely scenarios, but should refl ect combinations of the desirable and less desirable outcomes that will be a feature of most future trends. Scenarios provide multiple perspectives on key areas of uncertainty and allow the development of robust strategies that can deal with multiple outcomes.

The three scenarios are written from the perspective of a researcher in 2015 examining the current state of nanotechnology, what the key concerns are and the pathway that led to this point.

Scenario 1

Disaster Recovery

A lack of regulation resulted in a major accident. Public concern about nanotechnology is high and technology development is slow and cautious.

Scenario 2

Now We're Talking

Strong regulation and accountability systems are in place. The technology has been shaped by societal needs and strong health and safety concerns.

Scenario 3

Powering Ahead

Scientific progress has been faster than expected and nanotechnology is making a real impact, particularly in energy conversion and storage.

SCENARIO 1

Disaster Recovery

Public institutions have been slow to plan for the possibility of health or environmental risks related to nanotechnology and private enterprise has been reluctant to self-regulate. This lack of regulation contributed to a major accident at a manufacturing plant in Korea in 2012. Public concern about nanotechnologies escalated and a cautious approach to technology development was adopted. Although the technology is still being used and the science is still developing, the term nanotechnologies is used less, and the prefix nano has all-but disappeared.

The story so far...

2006

A campaign by a mass membership NGO to alert the public to the potential risks of nanotechnologies was undertaken. At the launch event a speech by a major respected public figure warned against "the insidious danger of meddling at the nanoscale". The campaign received little public support.

The final reports of public-funded projects to promote stakeholder dialogue on the social, environmental and economic risks and opportunities of nanotechnologies were produced, but received little attention.

2007

Nanotechnology-enabled consumer products went mainstream. Household paint that changed colour according to temperature was one such product. Another was anti-ageing cream. Later in the year an EU-funded study of the effects of nanoparticles on human health was published, showing some evidence for a negative effect. The report recommended more research to confirm the critical findings. A public opinion poll of European citizens showed that, among the minority that had heard of nanotechnology, most had positive associations with the term, though didn't necessarily trust public institutions to govern the application of the science effectively.

2008

An international symposium on nanotechnologies took place, at which agreement was reached about the need for a Global Framework on Emerging Technologies to regulate the production and use of nanotechnologies. Work started on developing the Framework. A brand of nanocoating for cars was recalled as it was found to peel off under extreme weather conditions and release nanoparticles into the environment.

2009

The combination of concerns around product safety and the lack of regulation meant that nanotechnology products were still peripheral in the marketplace. A major venture capital firm announced that it had embargoed all investment in nanotechnology-related products, citing a failure of the technology to deliver in the market as expected. Although a few other companies followed the lead, this decision was ridiculed by most in mainstream science. An editorial in 'Nature' magazine said the decision was "not only foolish, but dangerous."

2010

The UK Government publicly criticised the Global Framework on Emerging Technologies for moving too slowly and introduced its own, watered-down, guidelines. These were voluntary.

2011

Workers at a factory in Toulouse went on strike, refusing to work with nanoparticles following a number of medical complaints. Demonstrations spread across Europe. The number of occupational health related court cases increased.

A campaign by a major NGO was launched, calling for a moratorium on nanoscience and technologies until more was known about the health and environmental effects.

2012

In April, the process for delivering the Global Framework on Emerging Technologies broke down and efforts to create a level playing fi eld internationally were abandoned. A major explosion occurred at a plant on the outskirts of Seoul, which released several tonnes of nanoparticles into the environment.

2013

Routine monitoring of marine pollution in the Sea of Japan found high levels of nanopollutants in fish. This was traced to the Korean explosion. Further tests showed the particles in drinking water in Japan, leading to a public outcry.

It emerged that some athletes competing in the London Olympics 2012 were using nanotechnology-based performance enhancing drugs.

'Forbes' magazine stopped publishing its list of bestselling nanotechnology-related products.

2014

Residues of manufactured nanoparticles were discovered in Arctic sea ice.

A coherent EU regulatory framework for nanoscience and technology was finalised, based loosely on the UK guidelines.

2015

A consortium of European businesses published a report criticising the EU framework and committed to developing its own, stricter guidelines.

How Things have Changed Since 2006

- – – Proportion of scientists working in nanotechnology who feel that the media present a fair and balanced view of nanotechnology
- + Equality of access to nanotechnology-related products between the industrialised and developing world.
- – Proportion of nanotechnology-related patents originating in the European Union.
- – Public agreement with the statement that "nanotechnology on balance can make a signifi cant contribution to my quality of life".
- + Public sector funding for nanotechnology research and development
- + Penetration of nanotechnology-based products in food and packaging
- +++ Penetration of nanotechnology-based products in medical diagnostics
- ++ Penetration of nanotechnology products in the energy sector
- +++++++++ Number of nanotechnology pollution events
- + Number of nanotechnology-related patents fi led.

What's Selling Well?

Networked Earring

This is a powerful computer, small and light enough to be disguised as jewellery. It acts as the network hub for embedded microchips in clothing and interacts with local area networks providing a constant stream of communication between the user and their environment. It also connects to personal communication devices.

Nose Filter

This nose fi lter is a simple air-fi lter, capturing impurities using nanofi bremesh. It is worn inside the nose and is all-but invisible to the casual observer. The fi lter protects the wearer from allergenic spores and other particulate pollution.

Eezy-Spy

A cheap and easy-to-use surveillance service has been the surprise nanotechnology product success in 2015. Making use of the extensive networks linking embedded computers in most European towns and cities (and increasingly in the countryside), network operators provide tracking services to the general public. Tracking is only available with permission, and is increasingly being used to monitor the whereabouts of pets.

Longevity Cream

This popular anti-wrinkle and anti-bacterial cream is marketed as being able to extend the life of the user. The cream makes use of free nanoparticles, but the manufacturers have chosen not to market the cream as a product of nanotechnology.

Do-it-Yourself Medi-test

Available on supermarket shelves, this handy kit—employing lab-on-achip technology—tests the user for a range of medical conditions. It is easily networkable and comes with the optional extra of an automated on-line diagnosis and lifestyle advisory service.

What's Worrying Us?

Environmental Pollution

Environmental pollution is a key concern among the public. Although it is widely known that nanoparticles exist naturally and can be found everywhere, the Seoul explosion of 2012 put manufactured nanoparticles high on the 'undesirables' list—on a level with asbestos and nuclear radiation. Within this atmosphere of fear, a struggle is ensuing as to how legislation can be improved to prevent environmental damage. With centres of nanoscience now spread out around the world, achieving consensus is a major challenge.

Health and Safety

Health and safety for workers emerged as a key issue in 2011, when a spate of respiratory complaints and allergic reactions occurred among staff in a Toulouse factory. The symptoms were linked to the release of nanoparticles.

This has created a host of diffi culties for the nanotechnology industry. Today the sector remains exposed to litigation by staff and union action, its reputation is suffering and this is feeding through to difficulties in recruitment. Of greater concern, however, is the appearance of similar symptoms in consumers. Although this hasn't yet occurred, the possibility can't be discounted.

Privacy

The rapid development of ICT means that monitoring technology is increasingly pervasive and available for use by the public in a range of products and services. The convergence of medical diagnostic technology and ICT has introduced the possibility that people can, legitimately or otherwise, gain access to all sorts of personal data that might have a major impact on a person's employability, their ability to get medical insurance or to pay premiums.

More in Depth...

How is the Science Developing?

The lack of clarity on regulation, combined with public distrust of nanorelated products, has meant that progress in the science has been slower than aticipated and some products that seemed just around the corner a decade ago are still on the drawing board.

The trend now is for a greater emphasis in science and technology on understanding the potential risks of new developments, from a social, environmental and economic point of view, and a large amount of scientific funding is being diverted to precautionary studies of this type.

The chief exceptions to this are in the area of medical diagnostics, driven by military and space research funding, and ICT—both areas that avoid the use of free nanoparticles.

In general the term nanoscience has all but disappeared. The "sciences formerly known as nano" are no longer grouped together for funding purposes, and general nanoscience-focused journals and websites have almost all folded.

Is the Public Sector Investing?

Initially public sector funding was generous for nano-related research, on the basis that nanotechnology might provide solutions to many of the world's social and environmental problems. However, rising public concern, caused by events in Seoul and Toulouse, has led most European governments to adopt a more cautious approach and no-stringsattached public sector funding for basic nanotechnology is in decline.

Today public funding is increasingly focused on ensuring that the social, environmental and economic aspects of technologies are understood, and that there is an effective public debate about the role of science in society.

This approach is being encouraged at the international level, with the governments of the UK, Germany and Sweden putting pressure on other member states of the EU to restrict imports of nanotechnology-enabled products from less regulated markets by imposing high duties.

Is the Private Sector Investing?

From the mid-1990s onwards, there was a huge increase in private sector investment into research related to nanotechnologies. Today, however, private sector funding of science has reduced significantly. The public sector is funding the majority of basic scientific research at the nanoscale.

While some companies have increased their investment in studies (of variable quality) of the ethical, legal and social aspects of the technology used in products, there has also been a marked shift of investment out of the more regulated areas of Europe to other countries, particularly in south-east Asia, where the regulatory burden is lighter, or where enforcement of regulations is laxer.

How has Medical Diagnostics Changed?

Medical diagnostic technology has been in heavy demand from

the public health sector, but the greatest advances have been pioneered in research funded by the Indian Space Research Organisation and the National Space Development Agency of Japan, as well as the US military.

As early as 2009 astronauts were using devices for detailed self-diagnosis. In recent years pharmaceutical companies have marketed selfdiagnosis devices to consumers across the EU.

How has ICT Changed?

Because of advances in nanoscale engineering, computer microchip performance has taken enormous strides forward in the last few years.

Today, computers are more than a hundred times as powerful as they were in 2005. This has led to many new products and applications, with most products and many buildings containing tiny computers. Even animals, from pets to pests and endangered species, are commonly tracked and monitored in this way.

How Successful is Nanotechnology in the Market?

The Seoul disaster led to popular hostility to anything with the prefi x nano. As a result, products that rely on nanotechnology applications do not advertise the fact.

Today, the concept of nanotechnology-enabled products doesn't really exist in the way that was anticipated a generation ago, causing 'Forbes' magazine to stop publishing its annual list of bestselling products. Nonetheless, nanotechnology is still used, as expected, in new product development and the industry remains profi table, though a long way from realising the expectations of early "nanoenthusiasts".

Progress in applications reliant on the use of nanoparticles has slowed because of the amount of controversy over their effect on human health and the environment. Despite this, the cosmetic industry worldwide has continued to use nanoparticles in products, especially in developing country markets.

The most successful products avoid the use of free nanoparticles and are from one of two areas: ICT based on engineering at the nanoscale and the fi eld of medical diagnostics.

Products from both areas have found success in a range of different markets, including consumer markets.

How is Nanotechnology Regulated?

Despite early attempts, it has proven impossible to establish a level playing field globally for regulating the development of new technologies. Instead, we have a piecemeal approach. In Europe, we have a legal framework, finalised in 2014 and based on voluntary guidelines established by a joint private-public working group in the UK. The USA has a different set of laws, as do the other main producers of nanoproducts, such as China, South Korea, Indonesia and Brazil.

Recently, concerns have been raised that the framework isn't tight enough. This, and the suggestion that European borders are porous to products developed with less emphasis on safety and the environment, has led to individual companies developing their own codes of conduct that go beyond the regulatory requirement in a desire to be seen as responsible. However, the global picture is still very fragmented: it is diffi cult to see how one company's code of conduct aligns with another and to hold companies to account for their voluntary initiatives.

Have the Anticipated Risks Occurred?

There have been several major nanotechnology-related disasters in recent years, including most notably the explosion at a plant manufacturing nanotechnology-based drying agents in South Korea in 2012. This caused fi sh deaths on a large scale in the Sea of Japan, polluted drinking water along Japan's west coast and led to street demonstrations. At around the same time in Europe there was a sudden spate of court cases brought by workers in plants producing nanobased products, provoked by an increase in respiratory complaints and chronic allergies.

These complaints should have come as no surprise, as studies as far back as 2005 suggested that the release of nanoparticles into the environment might be dangerous. Further evidence emerged in the following years, but the results were always inconclusive which meant there was no concerted response from business or government.

How have Business Ethics Changed?

Initially, NGO campaigns to raise public awareness of the presence of nanotechnology in products largely failed, although the general antibusiness feel of NGO campaigns won support.

It is only in recent years, with several high profi le accidents pushing the issue into public consciousness, that business has truly begun to address the risks as well as the opportunities of nanotechnologies. Leading businesses are even going beyond what is legally required of them. All the signs are that these standards will raise the bar across the industry, at least in Europe.

How Effective is Public Debate?

Early expectations of radical social, economic and environmental benefits flowing from nanotechnology-enabled products have practically disappeared. Today, the dominant public discourse draws on a number of high profi le health or environmentrelated scandals. Many in science and industry feel that this is holding back scientifi c progress.

The media has adopted a sensationalist and adversarial approach to the issues and is perceived by science as ill-informed, obsessed with scandal and continually returning to a series of iconic failures. The substantial number of lower-impact successes has largely been ignored and level-headed debate informed by scientifi c method is hard to come by. In a recent interview one prominent scientist stated "there is not one journalist I would trust to deal with nanoscience in a mature and nuanced manner."

Likewise, in attempting to draw attention to risks, the NGO sector has missed the opportunity of supporting socially or environmentally benefi cial applications of the technology. NGOs have made the same mistake as the media in appearing to interpret nanotechnology as one monolithic entity, missing the fact that different nanotechnologies have different issues. This has played a part in forcing the prefix nano out of the public space, which has undermed NGOs' attempts to campaign on it.

In the past year there has been an improvement, though, and some companies are driving up standards of transparency and responsibility beyond the legal requirements. As a result, there are signs that a small amount of trust in nanoscience and nanotechnologies is being clawed back, but the popular assumption that nano is bad is still felt to be a major impediment to new product development and competing for research funds.

What does the Public Think?

Initially, campaigns by NGOs focused on the use of nanotechnology in products, but with little success as the use of nanotechnologies within supply chains was complicated and there was no shared defi nition of nanotechnology for governments to label or legislate on. Today, as poll upon poll has shown, the risks of nanotechnology are seen by the European public to outweigh the opportunities.

The very word nanotechnology has been demonised, and there is little appreciation of the fact that nanotechnology is a vast and diverse area.

The private sector has responded in part by abandoning the nano prefix. Although legitimate from a strictly scientifi c point of view, this has served to mask the use of nanotechnologies. As a result many of the ethical, legal and social issues that result from producing materials at the nanoscale are ignored.

Does Everyone have Equal Access?

Decades ago NGOs expressed concerns that the benefi ts of nano-based products would primarily be available only to affl uent consumers, opening up a "nanodivide" between the rich and poor. Today some of those concerns have been realised, due in part to the slow speed with which products have been developed and brought to market and because the anticipated economies of scale have not taken place.

The result is that nanotechnology-enabled products occupy the expensive end of product ranges. It is thought this is likely to change in the near future, however, as the geographical centre of production continues to shift eastward,and countries formerly thought of as developing begin to determine the sort of products that are released onto global markets. The untapped markets in these countries present innovative companies with a major opportunity. It is anticipated that as this opportunity is exploited, nanotechnology-enabled products will be available to a larger audience.

At present there are few organisations clamouring for private or public sector action to open up access to nanotechnology. If anything, despite the benefi ts that nanotechnologies could deliver, prominent NGOs are arguing that it is the poor, with lower awareness and access to information about products, who have less freedom to avoid potentially dangerous nanoparticles.

Have the Anticipated Opportunities Occurred?

The slower-than-expected development of nanotechnology has meant that fewer of the potentially transformative applications that were hyped in the early years have been launched onto the market. That said, there is obviously some belief in the potential opportunities of nanotechnologies, not only from an economic point of view, but also from social and environmental perspectives. If this did not persist, there would be even less money being invested in the science, given the known health risks. Not surprisingly, products that use applications not associated with the primary source of risk—nanoparticles—have proven to be more successful.

SCENARIO 2

Now we're Talking

Regulation of new technologies has been standardised internationally and strong accountability systems are in place, enabling transparent development of nanotechnology. Public sector incentives have directed research towards products that explicitly benefi t society, supported by public participation. Local stakeholder forums debate issues that arise from the use of technology (such as privacy) and make decisions for their local area. The strong regulatory regime, especially around issues of toxicity, has meant that health and safety risks are spotted early on and are well-managed. The focus on products that benefi t society and reduce environmental impact has paid off: growing resource stress means demand for these products is increasing around the world.

The Story so far...

2006

The European Commission developed a platform for dialogue between scientists, product developers, NGOs, consumer groups and others on the social and environmental aspects of nanotechnology. Early progress was made with some quick wins including:

- Funding allocated for The European Centre for Environment, Health, Safety and Toxicology (ECEHST)
- Moves to include training on the ethical, legal and social aspects of nanotechnology into all higher education courses
- An immediate review to establish the extent to which current regulation covers nanotechnology specific risks
- Development of a protocol for the assessment of risk and implementation of moratoriums, if necessary
- A requirement for all funding applications to be accompanied with a completed ethical, legal and social aspects (ELSA) assessment
- Education programmes and funding to support development of skills and mitigate anticipated skills shortage in Europe

Media workshops and other communication on the potential risks and benefits of nanotechnology were successful in galvanising a balanced and informed public discussion.

2007

The European Commission's Framework 7 research funding programme began. Research funds for the following seven years were directed towards "nanosciences, nanotechnologies, materials and new production technologies" and the extent to which they contribute to addressing European social, economic, environmental and industrial challenges.

2008

An OECD process for developing standards on nanoparticles commenced.

The ECEHST was opened. The centre identifi ed potentially harmful particles, provided guidance for regulation (e.g. where moratoriums were necessary) and advised on safety issues for workers and users.

2009

The OECD standards on nanoparticles were launched, hot on the heels of the Chinese standards.

An overhaul of the intellectual property/patenting system was announced.

2010

The first moratoriums were announced and a number of products were recalled, based on research from the ECEHST.

2011

The first local stakeholder debate took place after research linked a factory making metal oxide nanorods to cancer clusters.

2013

Privacy came to the forefront of the debate. Nanosensors tracked what people bought, where they went and even what they said. The media and civil rights groups branded this an infringement on civil liberty and the public took notice.

Stakeholder debates took place across Europe to discuss what was off limits. Clear signposts were required where the technology was in use, and products that used this type of surveillance technology were labelled accordingly.

2014

The nano tag was lost but this didn't mean the technology was not in abundant use. The science was everywhere, but not the name.

2015

BBC documentary 'Whatever happened to nanotechnology?' is

broadcast. The programme revisits 2006, the fears of the time and looks at developments of the past ten years. The programme takes viewers back to some of the more radical predictions from 2006, such as curing blindness.

It becomes clear throughout the documentary that the technology has not developed as fast as was predicted by some in 2006. On the other hand, none of the disasters predicted have materialised either. So on the whole, the documentary concludes, we are better off, the ground work has been laid and the future looks brighter.

How Things have Changed Since 2006

+ Proportion of scientists working in nanotechnology who feel that the media present a fair and balanced view of nanotechnology
− Equality of access to nanotechnology-related products between the industrialised and developing world.
no change : Proportion of nanotechnology-related patents originating in the European Union
− Public agreement with the statement that "nanotechnology on balance can make a significant contribution to my quality of life"
+ + Public sector funding for nanotechnology research and development
+ Penetration of nanotechnology-based products in food and packaging
+ Penetration of nanotechnology-based products in medical diagnostics
+ Penetration of nanotechnology products in the energy sector
no change : Number of nanotechnology pollution events
+ Number of nanotechnology-related patents filed

What's Selling Well?

Water Purifier

This nano-enabled water purifi er is now a mass-market product.

Nanomembranes effi ciently remove pollutants and bacteria. The biggest markets are in India and China. The African market is growing dramatically because of water stresses on the continent.

Rechargeable Batteries

High capacity nano-enabled rechargeable batteries were initially recalled because of concerns at the disposal phase. They are now available again after strict regulation controls have been established to control their collection for recycling.

All-in-one Household Protection

A high performance material cladding for building protection. The cladding is water-repellent, self-cleaning and provides heat insulation, thereby reducing household energy use.

Nano Food Packaging

These unique packaging systems use nanosensors to change the colour of the packaging when the food inside is no longer edible, and alerts a networked monitoring system. This helps the retail industry to guarantee product freshness and helps consumers to identify microbial contamination in the food they have bought. Similar features are available for home food-storage systems.

Eco-packaging

A recent breakthrough in biodegradable polymers, made stronger with the use of nanocomposites, means that they can now be used in more types of packaging (not just plastic bags). Such packaging can be recycled in composting facilities where the polymers, as well as nanoparticulate matter, become biological nutrients.

What's Worrying Us?

Missed Opportunities

Nanotechnology has been highly regulated. Potential risks have been flagged up early and a prudent approach has been developed which has enabled the continued success and acceptance of nanotechnology. Strong governance through the power and effectiveness of local stakeholder forums has contributed to nanotechnology's positive image in society.

Some, however, are complaining about the resultant lack of innovation and mounting bureaucracy—nanotechnology has not progressed as fast as was predicted in 2006.

Privacy

Debates around privacy issues rage on. In food packaging, obligatory surveillance of food during production and distribution through radio frequency identification (RFID) tags is intended to protect consumer health, but some customers feel eavesdropped by the technology. Without deactivation, the RFID nanosensor continues to monitor and save data in the consumer's fridge. This data can be recovered by unauthorized thirdparties once food-packaging is dumped.

Access

Although many products on the market have societal benefi ts, the high development costs (in part because of the highly rigorous approach to product testing and development) mean they remain out of reach to those that need them most. Belatedly, these issues are being addressed.

More in Depth...

How is the Science Developing?

Today, in Europe at least, new technology and applications must by law undergo rigorous tests to check for toxicity and other environmental impacts. This is in order to ensure that products will be completely safe throughout their life cycle. However, this has reduced the speed with which products are released onto the market. Public trust in nanotechnologies means that once products are approved, they are taken up quickly, and competition between companies is strong.

Is the Public Sector Investing?

Following an extensive consultation (with scientists, product developers, NGOs, consumer groups and others), which began in 2006, the EU government made a strong commitment to the development of nanotechnology—maximising potential benefi ts whilst doing everything in its power to minimise potential risks.

Funding was allocated at this early stage for the European Centre for EHS and Toxicology (ECEHST) following calls from all sides for further research on the potential toxicological impact of nanotechnologies on humans and the environment.

Government incentives have directed research towards products that benefi t society, particularly for use in developing countries, in line with the Millennium Development Goals. In addition to funding, governments play an important role in monitoring compliance and ensuring everyone is up to date on the latest requirements.

How has Global Economic Power Changed?

There is evidence that the relatively strong regulatory framework in the EU has driven investment overseas—where regulation is relatively weaker. However, stringent EU guidelines are driving standards up globally and the time-lag between the appearance of regulatory innovations in Europe and elsewhere appears to be decreasing. Indeed, China has been recognised as a world leader in the standardisation of nanotechnology since 2006, and in 2008 was a key player in initiating the successful OECD (Organisation for Economic Co-operation and Development) standardisation process.

How Healthy is the European Economy?

Increasing environmental pressures have raised awareness and consumer demand for carbon cutting and resource saving products as well as for renewable technologies. As a result, the early government moves towards encouraging developments in nanotechnology towards products that benefi t society has paid off. European companies are well positioned to respond to this increasing consumer demand, particularly from China and India—

both huge markets with equally huge environmental and resource pressures.

Is the World a Safer Place?

Low-level confl ict over resources such as natural gas and water, added to the continuing threat of global terrorism, has meant that security remains a top priority for all governments. Until recently, nanotechnology applications have been used widely to embed undetectable surveillance devices using RFID tags in the environment. Then, a few years ago, an NGO campaign sparked off a heated public debate over the increasing encroachment on civil liberty. Nowadays any area using this type of surveillance technology has to be clearly signposted or labelled and local stakeholder forums decide the extent to which the technology will be tolerated in their local area.

Is the Private Sector Investing?

Strong regulation has provided clarity for the private sector and focused the direction of its investment in nanotechnology.

Although returns on investment aren't excessive, investors are confi dent and fi nancial support for science and technology at the nanoscale has steadily increased over the years. Many private-public partnerships have been initiated, manufacturing and distributing products with maximum social and environmental benefit.

How is Nanotechnology Used with Food?

There is still a degree of public concern about the use of nanotechnology in food itself. But there have been major developments in food packaging, and the use of nanosensors that can detect contamination is now compulsory during the food manufacturing process, including its transport and distribution.

There have been a few cases in the press recently where nanosensors have malfunctioned, suggesting food is fresh that has in fact gone off—which have led to some high profi le class-action lawsuits. The courts granted those claims taking into account the food companies' product liability. There has been a back-to-basics movement reminding people also to "follow their noses" for the more traditional signs that food is not safe to eat.

How Successful is Nanotechnology in the Market?

Back in 2005, predictions of a "nanoboom" as more and more nanotechnology-related products hit the market were matched by fears of a "nanobubble", created by avid investment of venture capital, leading to a painful "nanoburst". This has, so far, been avoided, and nanotechnology is more a quiet success story than a consumerled frenzy.

This is largely down to the fact that the process of bringing new products to market is so carefully guided. Exhaustive testing is conducted on products in development stage and all potentially unsafe nanoparticles have been banned.

Nanotechnology is no longer used as a blanket term, as the technologies are so varied. As a consequence, companies no longer advertise the fact that certain products may be nanotechnology-related, and so it is diffi cult to track exactly how successful such products are. However, the environment, health and water sectors are all performing well and are intimately involved in nanotechnology. Affl uent consumers in particular are willing to pay more for products that reduce their environmental impact and their exposure to carbon taxes.

One unfortunate side-effect of the deliberative approach to product development is that the products tend to be more expensive. There is a danger as a result that the rich world benefi ts disproportionately from nanotechnology.

How is Nanotechnology Regulated?

A strict regulatory environment has evolved. International regulatory standards, promoted by the OECD, have been in place since 2010. Environmental and social impact assessments are now required for every new application that uses nanotechnology. Life cycle analysis is standard, analysing the impacts of each product from production, through use, to disposal. There is also a legal obligation to publish the results of all clinical studies from both the public and private sector.

Based on findings from the ECEHST, set up in 2013, there

have been a number of moratoriums put in place on certain applications of nanotechnolgy, which are then investigated further. Past moratoriums have included the use of nanotechnology in cosmetics and food supplements. The extensive body of research from ECEHST has revealed that, once we started looking for them and had the equipment to find them, nanoparticles were everywhere. As a result, there have been legislative shake-ups in many industries that have not traditionally been associated with nanotechnology, such as plastics, food manufacture and construction.

There is a central website resource from the ECEHST updated with all the information on the vast number of safety standards related to nanoparticles. This is mainly used by scientists and product developers but can be accessed by anyone for free.

EU and government funded multi-disciplinary teams that include representatives from NGOs, companies, regional governments and delegates from local stakeholder forums, advise on regulation and the direction of research funding.

How Effective is Public Debate?

Early mapping of key stakeholders enabled the European Commission to engage those with an interest in, those who might be affected by, or those who had a strong infl uence over, the development of nanotechnology—including scientists, product developers and other representatives from industry, NGOs, consumer groups, the media and academia. Effective dialogue at EU, national and regional levels has been key in directing nanotechnology towards more societal needs and building consensus over standardisation.

Educated through a series of high profile media workshops early on, the media has been key in providing informed and balanced information on nanotechnology and galvanising effective public debate.

Today stakeholder debates are regularly convened around issues of public interest. Although nanotechnology itself does not have a high enough profile to warrant specific debate, issues related to the impact of nanotechnology, such as privacy, do.

NGOs have also become much more targeted in their campaigning around specific issues—to great effect. There was a move away from blanket campaiging against all things nano once it became clear governments were helping to direct development towards meeting societal needs. Working in close partnership with ECEHST they remain critical in highlighting main concerns for further investigation.

What is Happening to the Environment?

Pressures on natural resources are increasing rapidly as the population continues to grow and as economic development in China, India and elsewhere continues apace. The focus for new technology development on innovating products that alleviate social and environmental problems has made a major contribution to reducing the impact of resource scarcity and nanotechnology, in particular, is seen as a critical means of continuing global economic growth within tightening environmental limits.

Have the Anticipated Risks Occurred?

Good regulation and strong governance have done much to prevent the scare stories of the early 2000s from materialising. Risks are spotted early (during research and development or lifecycle assessments) and dealt with quickly, before the product in question is released to market where it could pose a danger to the general public. There were some product recalls in the early days (nano food supplements, for example) following research fi ndings from the ECEHST.

Since then however, there have only been minor nanotechnology-related pollution or health events.

How have Business Ethics Changed?

For most of the last decade, the business community has been on the back foot when it comes to dealing with the ethics of nanotechnology. This is simply because the public sector in Europe has been so pro-active, and has pioneered regulation that guarantees a deliberative and precautionary approach to technology development. For example, legislation requires businesses to publish the results of clinical trials, and obliges businesses to participate in extensive stakeholder dialogue.

Business associations tend to complain about the amount of red tape and accuse governments of pushing up the price of goods through over-regulation. Individual companies, however, are keen to maintain a reputation for responsibility by publicly complying. A small number of leading companies even seek market differentiation by going beyond compliance, for example by applying high OECD standards in global affi liates.

What does the Public Think?

Today, the term nanotechnology is rarely used. Through open debate (for example, at local stakeholder forums) and education, the general public recognises that using the term nano as a prefi x to anything manipulated at the nanoscale is not particularly useful in understanding the benefits and impacts of the technology.

That said, the early identification of potential risks and the lack of headline grabbing horror stories mean that the overall impression of nanotechnologies is positive. It is also widely acknowledged that developments in technology (at the nanoscale) have enabled a lot of products to become available that have a social benefi t. Improvements in energy storage and water purifi cation in particular are seen to have had a positive impact.

Does Everyone have Equal Access?

An unintended consequence of the careful approach in taking new technologies to market has been to add a premium to nanotechnologyrelated products. Consultation and dialogue cost money, and it is eventually the consumer that pays the price for this. Although this hasn't affected the success of products in the market, it has contributed to an emerging "nanodivide" in Europe and the developing world. This is a matter of some concern in 2015, given that so many hopes for positive social and environmental benefi ts are pinned on nanotechnology.

Therefore, rather belatedly, signifi cant effort is going into developing new mechanisms to broaden access, although this poses many diffi culties. For example, some NGOs are entering into partnership with companies to deliver crucial products to "bottom of the pyramid" markets, often bringing in third-party companies based in the developing world. The continuing suggestion of public sector subsidy for the most important products, such as water purifi ers and air fi lters, is hotly debated.

In 2009, international patent law underwent a signifi cant overhaul in an attempt to prevent individual companies from wielding excessive market power and raising barriers to entry for new or smaller players, but fell short of limiting patents of novel materials. Today, there are renewed calls for further reform.

Have the Anticipated Opportunities Occurred?

Although the regulatory requirements for the development of nanotechnology have meant that opportunities have not been realised as speedily as was hoped, the strong steer towards benefi cial products means that any progress made tends to be in the right direction. More and more of the products that enter the marketplace benefit society. To realise their full potential, there is increasing recognition that more could be done to make these products accessible to all—particularly in developing countries where they are needed most.

SCENARIO 3

Powering Ahead

Scientific progress has been faster than expected and nanotechnologyrelated products are making a real impact on society and the economy. For example, there have been dramatic improvements in the efficiency of solar photovoltaic (PV) cells, with the result that applications expected to come into the market in the 2020s are already a reality. Long-term investments in fossil fuel resources are progressively losing value and new market entrants are growing quickly. The speed of change has left regulation behind. Although there has been discussion around the risks of novel materials, as far as public debate is concerned the benefi ts so far outweigh the risks.

The Story so far...

2007

Small efficient fuel cells entered the market and replaced

batteries in smaller electronic devices such as mobile phones and laptops.

Progress in this area drove research in other areas of fuel cell research and led to advances in larger fuel cell technology for transport use.

2008

There were dramatic improvements in PV—experimental solar cells were operating at 30 per cent effi ciency. Prices began to drop.

2009

Rapid developments occurred with the fi rst commercially available printable PV.

Governments across Europe struggled to keep up with the rapid pace of technological change. There was a lack of defi ned regulation. However, products were seen to have widely applicable benefi ts, so there were few objections.

European governments offered large subsidies to home-owners who invest in microgeneration.

2010

Printable PV was followed by spray-on solar.

There was an increasing shortage in engineers and researchers resulting in an increase in salaries.

There was a dramatic increase in the use of fuel cells in cars, at least ten years earlier than had been expected. Storage problems were solved by use of new composite materials and some houses were fi tted with fuel cells as power sources.

2011

The entrepreneur, scientist and author of 'The Microgeneration Revolution' died under suspicious circumstances. Inevitably, conspiracy theories connected this to certain energy companies being left behind by the new technology.

Many of the old energy giants lobbied hard against the decentralisation of energy production.

Greenpeace produced a report on resource use, which highlighted the limits of platinum availability and concerns about the lack of recycling of nanomaterials.

2012

A Nobel prize was awarded to the team responsible for developing cheap, effi cient spray-on solar cells.

Robotics started to kick off due to small, cheap and highly efficient batteries.

2013

The growth in nano-enabled products led to concerns over resource use and pollution. The recycling issue had still not been resolved.

The first major nanotechnology-related incident at a manufacturing plant highlighted the risks involved and forced a rethink from governments on regulation.

There was a worrying skills shortage in Europe.

2014

The rapid spread of spray-on solar cells led to a worldwide rise in renewable energy production. For the fi rst time there were signs that major reductions in CO^2 emissions might be achievable. The importance and timing of these developments cannot be overstated as atmospheric concentrations of CO^2 had reached 400ppm.

The religious right in the US scaled up its opposition to nanotechnology with a publication called 'The End of God's Children', which questioned the religious implications of the advancing science of human modifi cation.

2015

In 2015 the disruptive nature of the developments has become apparent as centralised energy production begins to fall dramatically. There is increasing unrest in countries that have no access to the technology and representatives are calling on governments and corporations to ensure wider distribution.

How Things have Changed Since 2006

No change: Proportion of scientists working in nanotechnology who feel that the media present a fair and balanced view of nanotechnology
– Equality of access to nanotechnology-related products between the industrialised and developing world
– Proportion of nanotechnology-related patents originating in the European Union
++++++ Public agreement with the statement that "nanotechnology on balance can make a significant contribution to my quality of life"
+++ Public sector funding for nanotechnology research and development
++ Penetration of nanotechnology-based products in food and packaging
+++++ Penetration of nanotechnology-based products in medical diagnostics
++++++++++ Penetration of nanotechnology products in the energy sector
+++ Number of nanotechnology pollution events
++++++ Number of nanotechnology-related patents fi led

What's Selling Well?

Fuel Cells for Transport

Although miniature fuel cells have been used in personal products for some years, it is only recently that improvements in catalysts and hydrogen storage have enabled the commercial rollout of fuel cells for use in transport.

Sola-shelter

Increasingly extreme weather, resource shortages and the resulting confl icts have increased the number of refugees and homeless.

Nanotechnologies have given rise to lightweight, strong materials with integrated photovoltaics. These have been used to construct a tough, re-useable, power-generating shelter.

The Walking Battery

Human clothing has integrated energy generation to supply increasingly power-hungry hi-tech personal devices.

Spray-on Photovoltaics

Although the pervasive nature of this application is often questioned, the benefi ts of quickly applied fl exible solar are huge. Any surface with a reasonable level of solar incidence can be turned to power generation.

Butlerbot

Advances in energy storage and computer processing have helped the development of semi-intelligent home assistants. Primarily built for simple tasks such as vacuuming, they have been adapted to help monitor energy consumption and provide basic surveillance.

What's Worrying Us?

Governance

Regulation and governance have been unable to keep up with the speed of technological development. While initially this did not have an impact, it is now evident that the energy divide and waste have to be addressed immediately.

Energy Divide

Although cheap, clean energy is increasingly available to citizens of developed nations and emerging economies, there is still an energy underclass who do not have the capital or infrastructure to benefit from the advances. This has lead to an increasingly bitter dispute about access and distribution of technology.

Resource Consumption

There has been a boom in cheap consumer goods and reliance on complex technological solutions has increased markedly. The question of recyclability remains unanswered and there is still enormous pressure on the planet's dwindling metal resources. Prices on the commodity markets are going up. For the countries rich in such resources, there is a temporary benefi t from mining revenues. But for other poorer regions, the situation threatens to push the cost of transformative technology beyond reach.

More in Depth...

How is the Science Developing?

Research in the energy fi eld increased dramatically from 2007 onwards, driven by an increase in investment from venture capitalists keen to develop opportunities in clean technology. This, combined with increasing awareness of environmental and social pressures, increased interest in PV research and stimulated several crucial breakthroughs in the application of nanotechnology.

Today the speed of progress in PV has exceeded the expectations of most scientists, who had not anticipated competitive fi nancial viability for solar technology until 2025 at the earliest. Flexible thin film solar is on sale and is predicted to have a dramatic impact on the energy market within the next ten years. Advances in PV have driven forward other areas of energy technology and a high number of energy-based applications using nanotechnology have been released onto the market. Super effi cient batteries, ultra-capacitors and fuel cells of a variety of sizes are becoming increasingly common in developed nations. There are still problems with long-term storage of energy from renewable sources, but further progress is expected in the energy sector with a high number of scientific papers on energy developments published recently.

There have been considerable advances in other applications of nanotechnologies, which have to some extent ridden on the back of the energy sector wave. The benefits of the advances in energy technology have also helped to allay fears of the impacts of nanotechnologies. There has been good progress in medical applications with advances in diagnosis and drug delivery.

Is the Public Sector Investing?

Riding on the wave of success, governments have continued to fund and support developments in nanotechnologies, although investment has started to wane in the face of take-up by the private sector. The military continues to invest heavily in a wide range of nanotechnology applications. It made considerable progress in energy conversion and storage in its pursuit for individual power supplies for troops and this has spun out clothing with integrated energy generation to provide power for electronic devices.

The success of energy related products has driven EU government commitment in other areas such as medicine and materials science.

How has Global Economic Power Changed?

China and India moved swiftly into the high tech energy market from 2007-2008 and have established market dominance in this area. Both countries have a high use of PV and have started ambitious projects to push the use of fuel cells in transport. The investment in science and engineering graduates in Asia in the fi rst decade of the 21st century has paid off and, unlike Europe and the US, the availability of a skilled workforce is not an issue. This has accelerated the shift in economic power and China is projected to overtake the US as the world's largest economy by 2020.

In Europe, despite calls from some politicians, business leaders and NGOs for increased investment in clean technology, the EU is still playing catch up with Asia. Thanks to the leadership of California and the massive investment there from venture capitalists, the US is in a slightly better position.

There are still a number of developing countries that have not been able to take advantage of the rapid development of technology due to lack of infrastructure and investment. There is growing demand for energy technology to be made more universally available.

Is the World a Safer Place?

The disruptive nature of developments in energy technology is becoming increasingly apparent and there has been a shift of power away from big oil since 2010. Many of the oil producing nations in the Middle East have invested considerable resources into large-scale solar developments and continue to sell energy to those countries without the infrastructure. But revenues are small compared to the golden era of mass oil consumption.

Even though the technology has the potential to provide cheap widelydistributed energy there is still an increasing divide between the rich and poor. This is due in part to the existence of a dramatic time lag between the introduction of the technology and its distribution to those who really need it. Large multinational companies from America and China hold many of the patents and there is increasing resentment from developing nations that cheap and plentiful energy is largely available only to the richer nations who now arguably need it less. To compound the impacts of the energy divide, many developing countries are suffering from resource shortages, with growing concerns over water scarcity and climate change. There is the threat of conflict over resources as the technology boom drives the demand for raw materials without rewarding the countries where the material originates.

Is the Private Sector Investing?

The investment community's faith in clean technologies has really paid off. The heavy investment in PV that started in 2006 has yielded swift advances and good returns on investment. As a result there is confi dence that nanotechnologies will continue to deliver in other areas, as well as energy. In 2007, private investment in nanotechnology-related research and development overtook the public sector contribution for the fi rst time and in 2015 is many times the size. In 2015 the PV market alone has revenues of $70 billion worldwide and is growing fast.

How has the Energy Market Changed?

New sources of energy have severely reduced the economic viability of older sources of generation such as nuclear, coal, gas and wind. PV began to compete seriously with these forms of generation in 2014. With pressure to develop renewable sources of energy, and concerns over security of supply, there has been rapid expansion in the roll-out of decentralised energy and microgeneration.

Recent developments are having a hugely disruptive effect on the energy market. Increased effi ciency in solar cells and improvements in batteries and fuel cell technology are changing the nature of energy production and distribution. With the advent of fl exible, economically viable solar, surface area is increasingly viewed as real estate and entrepreneurial individuals have started to sell prime locations on their property. The roll-out of fuel cells has lagged behind solar cells because of the delay between the introduction of the technology and the development of the necessary infrastructure.

How Successful is Nanotechnology in the Market?

The 'Forbes' top 10 list of nanotechnology products contains numerous energy devices and devices using nanotechnology-enabled power supplies. The developments in energy are the basis for hundreds of new product lines and nanotechnology enables many of these new products though local micro-power production. Used across the board, nanotechnology has become the disruptive technology that it initially promised to be. Products are often labelled "nanoenabled" as a marketing tool, even when the role of nanotechnology in their development has been relatively insignificant.

Are the Necessary Skills Available?

The rapid pace of technological development has begun to open up a skills gap, particularly in the EU. Although investment in science and education has provided an increase in science and engineering graduates, there is still some way to go to match countries such as India and China. There is considerable concern that this skills gap will begin to affect competitiveness and across the EU there are initiatives to encourage education in this area. In response the US and EU have opened up overseas academies to try to recruit science graduates from abroad.

How is Nanotechnology Regulated?

The regulatory environment has struggled to keep up with the pace of technological change and there is still a massive discrepancy between spending on developing the technology and researching the impacts in order to minimise them. Health and safety at the workplace and concerns over life-cycle impacts have recently made regulation a higher priority. As the technology has a high profile in society and a huge market value there is resistance from business interests to any additional red tape. Regardless of this pressure, governments are finally starting to react to the rapid development of technology and new legislation has been introduced in the last two years concentrating on health and safety, producer responsibility, end-of-life impact and release of nanomaterials into the environment.

What is Happening to the Environment?

Although many countries failed to meet their 2010 CO_2 reduction targets, there have been signifi cant reductions in CO_2 emissions across the EU. Progress since 2010 has been good and the EU is expected to meet the new target to reduce CO_2 emissions by 35 per cent by 2020. This was initially because of advances in energy effi ciency, but increasingly it is the result of the rapid uptake of solar energy generation by businesses and households. The prediction is that energy will no longer be a limiting factor on economic activity by 2050.

More of a concern is the availability of raw material resources. Although manufacturing at the nanoscale is improving efficiency, the breakthrough in the energy sector has led to an explosion in the number of personal hi-tech products. Because of the increasing complexity of these devices, and the fact that they are often embedded in other materials (as with solar clothing), recycling them effectively is extremely difficult. Not only does this create a waste problem, it also means that precious materials cannot be recovered and reused. This has placed upward pressure on commodity prices, with the likely effect that the costs will be passed on to consumers in the medium term.

Have the Anticipated Risks Occurred?

Dramatic scientifi c progress has led to greater speed of products to market and society has struggled to defi ne the implications of the advances in this technology.

As the new range of products entered the market between 2009 and 2012, several NGOs voiced concerns over the environmental and social impacts of the rapid introduction of new and complex materials, with claims that the world could experience the effects of "another asbestos", but on a bigger scale. However, the breakthroughs in energy technology have moved nanoproducts from the quirky to the real.

Genuinely useful products providing cheaper cleaner energy—leading, for example, to the phasing out of toxic materials in batteries—have meant that such warnings have gone largely unheeded. Until 2014 there was little discussion of the trade off between risk and benefit, even from the more environmentally minded NGOs, as the science began to deliver products that could help solve the energy crisis and reduce greenhouse gas emissions.

Initially the majority of the applications were deemed safe as the nanoparticles were fixed in the product. Problems first arose when the first wave of products were replaced. This increased concerns about material use, resource management and the diffi culties in recycling. Minor pollution events also raised the concern about the life cycle impacts of the products. In 2013, an accident occurred at a production site and staff suffered severe side effects from exposure to quantum dots. This incident reinvigorated NGO activity on the dangers of rushing products to market without research on the possible impacts.

The social impacts of these developments are only now starting to be realised. For example, there are concerns that reliance on hi-tech solutions will inevitably lead to problems of accessibility for the poor and potentially divert funding from simpler, more universally applicable solutions.

How have Business Ethics Changed?

Initially, those businesses involved in the development of energyproducts enjoyed unparalleled public support for delivering on a key environmental issue. These businesses heralded their products as truly sustainable and only a small number of NGOs questioned the long-term social and environmental impacts. The majority welcomed the advances in energy production and storage.

It was only once the first generation of devices reached the end of their life and the pervasive nature of spray-on solar became an issue, that the questions about material use and lifecycle impacts increased. NGOs began to raise the profile of the debate and those businesses involved had quickly to move on the wider, long-term issues associated with the life cycle of their products and major hurdles such as recycling. There are members of civil society who have accused business of ploughing ahead with progress without looking at possible impacts further down the road.

What does the Public Think?

In the years leading up to 2015, most public opinion polls showed an overwhelmingly positive response to nanotechnology.

Nanotechnologies are seen by most to be delivering an obvious societal benefit. Nanoenabled products are increasingly widely dispersed and accepted within civil society. Technology for the individual is beginning to take off in 2015 and there is a bullish view that nanoscience can deliver what was unimaginable just a decade ago.

However, debate has intensifi ed about the trade-off between rapid progress and the potential down-sides. Spray-on solar has raised awareness of the potential issues around how waste is dealt with when the product has reached the end of its useful life. The pervasive nature of the technology clearly exacerbates this problem substantially and there is increasing pressure for biodegradable alternatives to be developed.

There is increasing unease in religious circles about the path that the advances in technology are taking us. A multi-faith conference was held in early 2015 to challenge the technological progress and ask questions about the impacts on our society and our human nature. There is particular concern about the advances in the science of human modification.

How Effective is Public Debate?

The unexpected advances in energy technology have meant that until recently there was little call for debate on the value and trade offs in using nanotechnologies in energy products. However, today there are calls for a dialogue on a variety of topics from accessibility to waste impacts of solar cells. There is **growing alarm** around energy, social justice and the unfair distribution of technology. A number of NGOs who have continually called for dialogue and a more precautionary approach are now championing the idea that "the pursuit of technology for its own sake is a mistake" and society has to learn that "speed of development is not synonymous with progress".

The Future

Nanotechnology and AI/robotics, together with biotechnology, may well be on a convergent path. In 2001 the National Science Foundation held a large workshop to look at the implications of this convergence and the implications for human abilities and productivity. AI could be boosted by nanotechnology innovations in computing power. Applications of a future nanotechnology general assembler would require some AI and robotics innovations. Equally, nanotechnology may converge much sooner with biotechnology as it uses the tools and structures of biological systems to generate tiny machines. Although the prospect of general assemblers may be quite distant, self-replicating 'machines' that use the tools of biology—and look more like living things than machines—might be closer at hand through the convergence of bio- and nanotechnologies. Any creation that posed the prospect of being self-replicating would need to be handled with immense care to ensure environmental protection. Whether any of the technological futures being scoped out in laboratories are what our general public would like is a question that can only be answered by asking them. If those concerned with the development of new technologies, and nanotechnology in particular, are convinced that the benefits they hope to generate will withstand scrutiny they should have no concerns about engaging and winning public support.

Future File

Future of nanotechnology is potentially boundless if we can avoid the pitfalls. Some of the items that exist today were a topic of science fiction a decade ago and have the potential to transform our society very quickly.

Future Estimated Global Markets in Nanotechnology

Year	*Estimated global market*
2001	£31-55 billion
2005	£105 billion
2008	£500 billion
2010	£700 billion
2011-2015	Exceeds US$1 trillion (£0.6 trillion)

11

GRAND CHALLENGES

Society is in danger of squandering the powerful potential of nanotechnology due to a lack of clear information about its risks, conclude 14 top international scientists in a major paper published in the November 16th issue of the journal Nature. The paper, "Safe Handling of Nanotechnology," identifies Five Grand Challenges for research on nanotechnology risk that must be met if the technology is to reach its full promise.

The paper's lead author is Project on Emerging Nanotechnologies Chief Science Advisor Andrew Maynard. The co-authors are among the world's foremost nanotechnology risk and applications researchers from universities, government, and industry in the United States and Europe.

"The spectre of possible harm—whether real or imagined— threatens to slow the development of nanotechnology unless sound, independent and authoritative information is developed on what the risks are, and how to avoid them," Maynard and his co-authors write.

"We are running out of time to 'get it right.' Last year, more than \$32 billion in products containing nano-materials were sold globally. By 2014, Lux Research estimates that \$2.6 trillion in manufactured goods will incorporate nanotechnology," asserts Maynard. "If the public loses confidence in the commitment—of governments, business, and the science community—to conduct sound and systematic research into possible risks, then the enormous potential of nanotechnology will be squandered. We cannot let that happen."

"Fears over the possible dangers of some nanotechnologies may be exaggerated, but they are not necessarily unfounded," say the authors. "Recent studies examining the toxicity of engineered nanomaterials in cell cultures and animals have shown that size, surface area, surface chemistry, solubility and possibly shape, all play a role in determining the potential for nanomaterials to cause harm."

The paper outlines Five Grand Challenges to "stimulate research that is imaginative, innovative, timely and above all relevant to the safety of nanotechnology." They include the development of:

1. Instruments to assess environmental exposure to nanomaterials,
2. Methods to evaluate the toxicity of nanomaterials,
3. Models for predicting the potential impact of new, engineered nanomaterials,
4. Ways of evaluating the impact of nanomaterials across their life cycle,
5. Strategic programs to enable risk-focused research

Within the Five Grand Challenges, the authors set specific targets to achieve within specific timeframes. These include developing a "universal aerosol sampler" for measuring exposure to airborne nanomaterials, assessing whether fiber-shaped nanoparticles present a unique health hazard, and establishing ways of engineering nanomaterials that are "safe-by-design."

"It is generally accepted that, in principle, some nanomaterials may have the potential to cause harm to people and the environment," according to the authors. "Yet research into understanding, managing, and preventing risk often has a low priority in the competitive worlds of intellectual property, research funding and technology development."

"Ultimately, this is not just a question about nanotechnology," says Maynard. "It is about whether governments, industry and scientists around the world are willing to make safe nanotechnology a priority. Are they willing devote the resources necessary to develop a comprehensive research strategy, and to work together with some urgency to implement and enable the technology to be safely applied?"

Maynard and his co-authors conclude that "if the global research community can take advantage [of the research opportunities before us] and rise to the challenges we have set, then we can surely look forward to the advent of safe nanotechnologies."

Difficult Problems in Nanotechnology

1. Precise and arbitrary manipulation and positioning of large numbers of nanostructures.
2. The development of improved techniques for multi-scale modeling (i.e., unified approaches for the modeling across multiple scales of extended systems of nanostructures).
3. The design, fabrication, and demonstration of an extended nanocomputer system that is integrated on the molecular scale (i.e., the nanometer scale), including both an ultra-dense nanoprocessor and an ultra-dense nanomemory array.
4. Development of techniques for imaging atomic-scale features in real time under a wide range of conditions.
5. Better understanding of the health issues related to nanoparticles and other nanostructures.
6. Bulk synthesis or bulk separation of carbon nanotubes with controlled chirality.
7. Improved theories for understanding and predicting physical processes (chemical reactions, atomic transport, crystal structures, etc.) at nanometer length scales.

Moral Implications

Society also limits technology. And with a technology as revolutionary as nanotechnology, people may place far more restrictions on this.

How will they spend their time? With nano machines doing almost everything, only creative work will remain. And if one can't be creative enough, what will she/he do? If one acquires material possessions free of charge, does she/he have to work? If there's no work, no jobs, what will the government tax? Nanotechnology's impact on the political, economic, and social cover of society could be very real.

True, nanotechnology conjures up visions of utopian existence, but they could also give rise to a lot more risks to our lives. For instance, nanotech can enable unlimited supply of weapons. If let off into the environment, nano machines could reduce all matter to dust, the 'grey-goo' fear. Possible destruction of the biosphere will certainly lead to control of the technology or even a ban.

And most importantly, if you can make anything easily and cheaply, what do you choose to make? Society will place restrictions on what is done based on common social values.

The ultimate limit on nanotechnology may be that some things just cannot be done because they are too complex.

FREQUENTLY ASKED QUESTIONS

Frequently Asked Questions related to Nanotechnd ogy (FAQ)

1. *What is nanotechnology?*

Nanotechnology is science and engineering at the scale of atoms and molecules. It is the manipulation and use of materials and devices so tiny that nothing can be built any smaller. Materials at the nanoscale are typically between 0.1 and 100 nanometres (nm) in size –1 nm is one billionth of a metre (10^{-9} m).

2. *How did the idea of nanotechnology emerge?*

Work done early this century clarified the nature of matter and atoms, showing how atoms combine. Research by chemists in the 1950s showed the workings of natural molecular machines. In a 1959 talk, physicist Richard Feynman proposed that tiny robots might be able to build chemical substances. At MIT in 1977, as an outgrowth of studies of naturally-occurring molecular machines, Eric Drexler developed the essentials of the current concept of nanotechnology. These ideas were first presented in a scientific journal in 1981, and in a book in 1986. He taught the first course on the subject at Stanford University in 1988.

3. *Will developing nanotechnology require new scientific discoveries?*

The basic properties of atoms and molecules are already well understood, though routine research will be part of the development process. The existence of molecular machines in nature shows that machines at that scale are physically possible. No new fundamental science is needed; nanotechnology will be an engineering advance. This makes it foreseeable, unlike future scientific discoveries.

4. *How is nanotech different from biotech?*

Biotech can be thought of as a subset of nanotech—"nature's nanotechnology." Biotech uses the molecular structures, devices, and systems found in plants and animals to create new molecular products. Nanotech is more general, not being limited to existing natural structures, devices, and systems, and instead designing and building new, non-biological ones. These can be quite different: harder, stronger, tougher, and able to survive a dry or hot environment, unlike biology. For example, nanotech products can be used to build an automobile or spacecraft.

5. *Where can we see nanotechnology in consumer devices and other applications right now ?*

According to lists compiled by the US government's National Nanotechnology Initiative, nanotechnology can already be seen around us today in:

- Computer hard drives
- Car parts and catalytic converters
- Scratch- and wear-resistant paints and coatings
- Sunscreens (titanium dioxide nanoparticles are transparent, yet absorb UV light at the same time) and lipsticks
- Longer lasting tennis balls, and hardwearing yet lightweight tennis racquets
- Metal cutting tools
- Antibacterial bandages incorporating silver nanoparticles
- Anti-static packaging for sensitive electronic equipment
- Nanofilm-coated "self-cleaning" windows and
- Stain-resistant fabrics

6. *Are there any real world examples showing the promise of nanotechnology?*

Yes. The field of nanotechnology has already yielded specific products and proofs of principle demonstrated to be of value in clinical applications:

Liposomes, which are first generation nanoscale devices, are being used as drug delivery vehicles in several products. For example, liposomal amphotericin B is used to treat fungal infections often associated with aggressive anticancer treatment and liposomal doxorubicin is used to treat some forms of cancer.

Another recent example is work done by researchers at Massachusetts General Hospital, led by Ralph Weissleder, M.D., Ph.D., which has shown that nanoparticulate iron oxide particles can be used with magnetic resonance imaging (MRI) to accurately detect metastatic lesions in lymph nodes without surgery.

In May 2004, two companies (American Pharmaceutical Partners and American BioScience) announced that the FDA accepted the filing of a New Drug Application (NDA) for a nanoparticulate formulation of the anticancer compound taxol to treat advanced stage breast cancer. Earlier-stage research has shown that nanoparticulate sensors can detect the cell death that occurs when a cancer cell succumbs to the effects of an anticancer drug. Such sensors would be of great value to clinicians, who would no longer have to wait months to determine if a cancer therapy is working before switching a patient to a course of therapy. In addition, such a sensor could greatly accelerate clinical trials of new anticancer agents, again by providing very early signals of efficacy.

7. *What enabling technologies could be useful in developing nanotechnology ?*

Enabling technologies that could be useful in the development of nanotechnology include research in proximal probes such as STMs and AFMs, protein design and molecular design of molecules containing large quantities of constituent atoms (especially carbon based), self-assembling molecules and molecule design, mechanosynthetic chemistry, and the continuing development of more powerful computer systems and enhanced chemical modeling packages.

8. *In which areas can the use of nanotechnology prove beneficial ?*

The major areas of application right now are medical and pharmaceuticals, information technology, chemicals and advanced materials. However, if we look at this from an Indian perspective, textiles are far more significant.

The major applications would be in chemicals (better catalysts to improve throughput, the use of nano particles in composite materials) and textiles (to make cloth stain and crease resistant).

Nanotechnology will also affect packaging. For an agriculturally rich nation like India, getting fresh food to outside markets in peak condition will be very a much a function of how it is packaged—whether we are talking about 'smart' packaging, for example, the use of gas permeable layers that can expel undesirable gases such as oxygen while replacing it with nitrogen, or simply being able to detect a consignment of milk that has been left out on the sun en route to the supermarket.

9. *What are nanoscale biomedical devices ?*

Nanoscale devices are the same size as many important biological objects, and therefore can be used to "see" biological activity that the naked eye cannot. They can perform tasks inside the body that would otherwise not be possible. For example, nanoscale devices smaller than 50 nanometers can easily enter most cells, while those smaller than 20 nanometers can move through the walls of blood vessels. As a result, nanoscale devices can readily interact with molecules on both the cell surface and within the cell, often in ways that do not alter the behavior of those molecules.

Nanoscale devices are smaller than human cells, which are 10,000 to 20,000 nanometers in diameter, and cellular components such as mitochondria that are inside cells. Nanoscale devices are similar in size to large biomolecules such as enzymes and receptors—hemoglobin, for example, is approximately five nanometers in diameter, while a cell's wall is around six nanometers thick. Larger devices, such as microfluidic chips, are also being developed with nanoscale components for advanced diagnostics applications.

10. *Why is nanotechnology an important tool for cancer research?*

There are several reasons that nanotechnology could help transform cancer research and clinical approaches to cancer care:

Most biological processes, including those processes leading to cancer, occur at the nanoscale. For cancer researchers, the ability of nanoscale devices to easily access the interior of a living cell affords the opportunity for unprecedented gains on both clinical and basic research frontiers.

The ability to simultaneously interact with multiple critical proteins and nucleic acids at the molecular level will provide a better understanding of the complex regulatory and signaling patterns that govern the behavior of cells in their normal state as well as the transformation into malignant cells.

Nanotechnology provides a platform for integrating research in proteomics the study of the structure and function of proteins, including the way they work and interact with each other inside cells with other scientific investigations into the molecular nature of cancer.

11. *What Are the Main Advantages of Molecular Building Blocks?*

A huge technology base for molecular construction already exists. Tools originally developed by biochemists and biotechnologists to deal with molecular machines found in nature can be redirected to make new molecular machines. The expertise built up by chemists in more than a century of steady progress will be crucial in molecular design and construction. Both disciplines routinely handle molecules by the billions and get them to form patterns by self-assembly. Biochemists, in particular, can begin by copying designs from nature.

Molecular building-block strategies could work together with proximal probe strategies, or could replace them, jumping directly to the construction of large numbers of molecular machines. Either way, protein molecules are likely to play a central role, as they do in nature.

12. *How will nanotechnology be used in computation?*

Assembler-based manufacturing will enable the construction of extremely small computers. The equivalent of a modern mainframe computer could fit into a cubic micron, a volume far smaller than that of a single human cell. Once such nanocomputers have been designed and the technology is in hand, building them will be inexpensive, enabling us to use many of them at once. A laptop computer could then have more power than all the computers in the world today put together

13. *How will nanotechnology be used to benefit the environment?*

By giving thorough control of matter, nanotechnology will enable us to end chemical pollution: any waste atoms could be recycled, since they could be kept under control. By reducing the cost of environmental cleanup and freeing land area from industrial uses, assembler-built products should aid environmental restoration. For example, even the immense tonnage of excess carbon dioxide in the atmosphere—a chief cause of greenhouse warming—could be swiftly and economically removed.

14. *What are potential applications of nanotechnology in occupational safety and health?*

Nanotechnology holds great promise for society, and occupational safety and health is no exception. Engineered nanomaterials may support the development of high performance filter media, respirators, coatings in non-soiling/dust-repellant/self-cleaning clothes, fillers for noise absorption materials, fire retardants, protective screens for prevention of roof falls and curtains for ventilation control in mines, catalysts for emissions reduction, and clean-up of pollutants and hazardous substances. Nanotechnology-based sensors and communication devices may help in handling emergencies and in empowering workers to take preventative steps to reduce their exposure to risk of injury. The smallness of their size coupled with wireless technology may facilitate development of wearable sensors and systems for real time occupational safety and health management. Nanotechnology-based fuel cells, lab-on-chip analyzers and opto-electronic devices all have the potential to be useful in the safe, healthy and efficient design of work itself.

15. *Could human body fluids get inside the nanorobot?*

From a medical standpoint, it makes sense to regard the nanorobot as having two spaces which should be considered separatelyits interior and its exterior. It is true that the nanorobot exterior will be exposed to the diverse chemical brew that makes up our human biochemistry. But the interior of the nanorobot may be a highly controlled environment, possibly a vacuum, into which external liquids cannot normally intrude.

Of course it may often be necessary for a nanorobot to import external fluids in a controlled manner for chemical analysis or other purposes. But the important thing is that the device will be watertight and airtight. Body fluids cannot get into the interior of the device, unless these fluids are purposely pumped in for some specific reaso

16. *What would a typical nanorobot look like ?*

It is impossible to say exactly what a generic nanorobot would look like. Nanorobots intended to travel through the bloodstream to their target will probably be 500-3000 nanometers (1 nanometer = 10-9 meter) in characteristic dimension. Nonbloodborne tissue-traversing nanorobots might be as large as 50-100 microns, and alimentary or bronchial-traveling nanorobots may be even larger still. Each species of medical nanorobot will be designed to accomplish a specific task, and many shapes and sizes are possible.

Finally, and perhaps most importantly, no actual working nanorobot has yet been built. Many theoretical designs have been proposed that look good on paper, but these preliminary designs could change significantly after the necessary research, development and testing has been completed.

17. *I have read that self-replicating nano-robots of the future could run wild, turning all life on Earth into a dead "grey goo". Is this likely, or even plausible?*

Most scientists think this a pretty improbable scenario.

Even futurist K Eric Drexler, who came up with the idea in the first place (in his 1986 book Engines of Creation) has since recanted, and Greenpeace UK's chief scientist describes grey goo as a "distraction from the real issues".

"Self-replicating, mechanical nano-robots are simply not possible in our world," according to Richard Smalley a chemist at Rice University in Houston, US. Smalley, co-discoverer of carbon buckyballs, argues that the chemistry just does not add up.

Simply shrinking large-scale machines and components down to the nano realm probably would not work, as the laws of nature are much different at small scales. For example: water becomes as thick as treacle, Brownian motion buffets things all over the place, and particles become much more "sticky" and attracted to one other.

18. *What is the biggest challenges for nanotechnology?*

The perception problem. Because people cannot understand nano, this technology is a little removed from the public domain.

Companies and governments need to come together and make people aware of this technology.

19. *Where is nanotechnology headed now ?*

Within the next 10 years, we will be seeing nano science in universities to a much greater degree than ever before. But most importantly, we will be seeing its applications in water treatment, drug delivery, the preservation of food and in the power industry.

20. *Does nanotechnology offer an entrepreneur opportunities?*

The nano market is estimated to be worth $3 trillion. One can be tempted to conclude that there is enough opportunity for everyone to do something nano and expect big returns. But we need to look beyond those numbers and identify the economics of our set up.

From a business point of view, we want to make better things that are cost effective but, at the same time, profitable. Nanotechnology when applied in the correct form can result in bigger gains but has the potential to backfire as well.

Nanotechnology has resulted in the discovery of new materials which can be used in a range of industries, from manufacturing to healthcare.

So an entrepreneur who wants to invest in nanotechnology must identify what properties of the new material he can exploit. One mistake that many start ups in nanotechnology make is that they invest again in the research of the technology.

A smart businessman will look at the available technology and tweak it to his benefit. Nano start ups can take cues from the world of information technology. It took companies like IBM to set up the basic framework of computers, servers and networks while companies like Fed Ex, United Parcel Service (UPS) and Amazon used it to their benefit.

21. *What effects will nanotechnology have on the economy?*

It will fundamentally revolutionize most industries, and has been compared in importance to humanity's taming of fire. The Industrial Revolution pales in comparison. Because assemblers will be able to build copies of themselves quickly, using inexpensive materials, little energy, and no human labor, a single assembler can be used to make billions. Once we have software to program assemblers to make consumer goods, each household could use an assembler system to produce goods cheaply and quickly. Manufacturing, mining, transportation, and other industries will change radically. Individuals will be able to make at home much of what they need, reducing the need to transport goods. This should encourage decentralization.

22. *Where India stands in nanotech ?*

India has started ten units of nanoscience and seven centres of nanotechnology with teams working in the area at the IITs, the IISc and the Jawaharlal Nehru Centre for Advanced Scientific Research (JN-CASR), among others. The Government has recently decided to invest much more in nanoscience and technology over the next five years.

India need to attract young people to the field as well as create facilities. So far, funding by Indian Govt. is very small. Other countries have invested millions of dollars.

India can, start the large-scale production of nanosensors right away, but scientist need industrial support for that.

The next three to four years should see India making several break-throughs in the field.

23. *What can I do to help?*

The challenges brought by advanced nanotechnology will have to be addressed by a diverse collection of people and organizations. No single approach will solve all problems or address all needs. The only answer is a collective answer, and that will demand an unprecedented collaboration—a network of leaders in science, technology, business, government, and NGOs. You can contact me at visionespkota@yahoo.com

GLOSSARY

Atomic Force Microscopy (AFM)

A type of scanning probe microscopy that maps the topography of an interface by scanning a force sensor over the interface.

Autoproductivity

The ability of a system, under external control, to automatically produce an identical copy of itself.

Buckyball

Round (convex) fullerenes ranging in size from 20 to over 500 carbon atoms.

Bottom Up

Building larger objects from smaller building blocks. Nanotechnology seeks to use atoms and molecules as those building blocks. The advantage of bottom-up design is that the covalent bonds holding together a single molecule are far stronger than the weak

Brownian Motion

Motion of a particle in a fluid owing to thermal agitation, observed in 1827 by Robert Brown. (Originally thought to be caused by vital force, Brownian motion in fact plays a vital role in the assembly and activity of the molecular structures of life.

Biosensor

An analytical tool consisting of biologically active material used in close conjunction with a device that will convert a biochemical signal into a quantifiable electrical signal.

Cantilever

A projecting structure that is supported at one end and carries a load at the other end or along its length, like a diving board

Core/Shell Nanoparticles

Nano-spheres of one metal coated with a continuous shell of another.

Chemical Vapour Deposition (CVD)

A technique used to deposit coatings, where chemicals are first vaporized, and then applied using an inert carrier gas such as nitrogen

Cognotechnology

Convergence of nanotech, biotech and IT, for remote brain sensing and mind control

Computational Nanotechnology

Permits the modeling and simulation of complex nanometer-scale structures. The predictive and analytical power of computation is critical to success in nanotechnology: nature required several hundred million years to evolve a functional "wet" nanotechnology; the insight provided by computation should allow us to reduce the development time of a working "dry" nanotechnology to a few decades, and it will have a major impact on the "wet" side as well

Convergent assembly

A process of fastening small parts to obtain larger parts, then fastening those to make still larger parts, and so on; convergent assembly can be used to build a product from many, much smaller, components

CMOS

An acronym for complementary metal-oxide-semiconductor, as in CMOS transistor and CMOS logic

Cell pharmacology

Delivery of drugs by medical nanomachines to exact locations in the body

Dip Pen Nanolithography (DPN)

A method for nanoscale patterning of surfaces by the transfer of a material from the tip of a scanning probe microscope onto the surface.

Dendrimers

From the Greek word dendratree, a dendrimer is polymer that branches. .a tiny molecular structure that interacts with cells, enabling scientists to probe, diagnose, cure or manipulate them on a nanoscale." Invented by Professor Donald Tomalia from Central Michigan University.

DNA Chip

A purpose built microchip used to identify mutations or alterations in a gene's DNA

Dry Nanotechnology

Derives from surface science and physical chemistry, focuses on fabrication of structures in carbon (e.g. fullerenes and nanotubes), silicon, and other inorganic materials. Unlike the "wet" technology, "dry" techniques admit use of metals and semiconductors. The active conduction electrons of these materials make them too reactive to operate in a "wet" environment, but these same electrons provide the physical properties that make "dry" nanostructures promising as electronic, magnetic, and optical devices. Another objective is to develop "dry" structures that possess some of the same attributes of the self-assembly that the wet ones exhibit.

Electron Beam Lithography (EBL)

A method of fabricating sub-micron and nanoscale features by exposing electrically sensitive surfaces to an electron beam.

The method is similar to photolithography, but uses electrons rather than photons. Since the wavelength of an electron is far smaller than that of a photon, diffraction is not a limit to the resolution. While EBL is more expensive and less parallel than photolithography, its resolution is higher and it is frequently used to create photolithographic masks.

Ecosystem protector

A nanomachine for mechanically removing selected imported species from an ecosystem to protect native species.

Femtotechnology

The art of manipulating materials on the scale of elementary particles (leptons, hadrons, and quarks). The next step smaller after picotechnology, which is the next step smaller after nanotechnology

Fullerene

A class of cage-like carbon compounds composed of fused, pentagonal and/or hexagonal sp^2 carbon rings.

Grey Goo

A scary concept dreamed up by Erik K Drexler whereby tiny assemblers, or molecular machines, that are capable of making copies of themselves, are let loose and proceed to replicate uncontrollably, consuming everything in their path and turning it into a grey goo.

Green Goo

Nanomachines or bio-engineered organisms used for population control of humans, either by governments or eco-terrorist groups. Would most probably work by sterilizing people through otherwise harmless infections.

LCD

Liquid Crystal Display is the predominant technology used in flat panel displays. The principle that makes the display work is: A crystal is alignment can be altered with an electric current. If the crystal's lined up one way then it will allow the light waves to pass through a polarized filter, but if the electric current alters the crystal's alignment, it will guide light so that the polarized filter blocks the light. By densely packing red, blue and green light emitting crystals next to each other on a sheet (called a substrate), one can create a full color display. The great thing about LCD is that the crystals can be packed together closely, allowing for a higher-resolution, finer-detail display.

Molecular Nanotechnology (MNT)

Thorough, inexpensive control of the structure of matter based on molecule-by-molecule control of products and byproducts; the products and processes of molecular manufacturing, including molecular machinery

Magnetic Force Microscopy (MFM)

A method for observing local magnetic fields near a surface by scanning the surface with a magnetic probe

Micro-Electromechanical Systems (MEMS)

"The fabrication or micro-machining of materials to make stationary and moving structures, devices and systems of a nominal size scale from a few centimeters to a few micrometers."

Moore's Law

An emperical trend in the microelectronics industry for the number of circuits per chip to double roughly every 18 months.

Molecular electronics

Any system with atomically precise electronic devices of nanometer dimensions, especially if made of discrete molecular parts rather than the continuous materials found in today's semiconductor devices

Molecular manipulator

A device combining a proximal probe mechanism for atomically precise positioning with a molecule binding site on the tip; can serve as the basis for building complex structures by positional synthesis

Molecular recognition

A chemical term referring to processes in which molecules adhere in a highly specific way, forming a larger structure; an enabling technology for nanotechnology

Nanomachine

Also known as 'nanites', nanomachines are mechanical devices so small that the parts are single molecules.

NEMS (NanoElectroMechanical Systems)

Systems consisting of integrated electromechanical actuators of nanometer-scale dimensions.

Nanoimprint Lithography (NIL)

A method of lithography that uses a mold, or mechanical force (embossing) to pattern a resist.

Nuclear Magnetic Resonance (NMR)

An analytical method that can detect subatomic and structural information of molecules by measuring the adsorption of radio-frequency electromagnetic radiation by nuclei under the influence of a magnetic field.

Nanoarray

An ultra-sensitve, ultra-miniaturized array for biomolecular analysis. BioForce Nanosciences' Nanoarrays utilize approximately 1/10,000th of the surface area occupied by a conventional microarray, and over 1,500 nanoarray spots can be placed in the area occupied by a single microarray domain

Nanobeads

Polymer beads with diameters of between 0.1 to 10 micrometers. Also called nanodots, nanocrystals and quantum beads. Impregnating fluorescent crystal chips into these beads allows simultaneous measurement of thousands of biological interactions, a stepping stone for breakthroughs in the diagnosis and treatment of disease.

Nanobiotechnology

Applying the tools and processes of MNT to build devices for studying biosystems, in order to learn from biology how to create better nanoscale devices. Should hasten the creation of useful micro devices that mimic living biological systems.

Nanochips

We are approaching the limits of standard microchip technology; thus, the "nanochip"—a next-smaller microchip. They are also a next-gen device for mass storage, of significantly higher density, with greater speed, and much lower cost

Nanocomputer

A computer made from components (mechanical, electronic, or otherwise) built at the nanometer scale. These computers could be many orders-of-magnititude faster than today's, which enables software to take proportional leaps

Nano Cubic Technology

An ultra-thin layer coating that results in higher resolution for recording digital data, ultra-low noise and high signal-to-noise ratios that are ideal for magneto-resistive (MR) heads. It is capable of catapulting data cartridge and digital videotape to one-terabyte native (uncompressed) capacities and floppy disk capacities to three gigabytes. To help visualize the potential, 1TB can store up to 200 two-hour movies

Nanoelectronics

Electronics on a nanometer scale, whether made by current techniques or nanotechnology; includes both molecular electronics and nanoscale devices resembling today's semiconductor devices

Nanogate

A device that precisely meters the flow of tiny amounts of fluid. Precise control of the flow restriction is accomplished by deflecting a highly polished cantilevered plate. The opening is adjustable on a sub-nanometer scale, limited by the roughness of the polished plates. Thus, the Nanogate is an Ultra Surface Finish Effect Mechanism (USFEM). The Nanogate can be fabricated on a macro- , meso- or micro- (MEMs) scale

Nanorods or Carbon Nanorods

Formed from multi-wall carbon nanotubes. Another nanoscale material with unique and promising physical properties, such that may yield improvements in high-density data storage, and allow for cheaper flexible solar cells.

Nanoshells

Nanoscale metal spheres, which can absorb or scatter light at virtually any wavelength. The nanoshells act as an amazingly versatile optical component on the nanometer scale: they may provide a whole new approach to optical materials and components.

Nanofactory

A self-contained macroscale manufacturing system, consisting of many molecular manufacturing systems feeding a convergent assembly system

Nanometer

One billionth of a meter; approximately the length of three to six atoms placed side-by-side, or the width of a single strand of DNA; the thickness of a human hair is between 50,000 and 100,000 nanometers.

Nanocomposite

Nanomaterials that combine one or more separate components in order to obtain the best properties of each component (composite). In nanocomposite, nanoparticles (clay, metal, carbon nanotubes) act as fillers in a matrix, usually polymer matrix.

Nanomaterials definition

Nanomaterials can be defined as materials which have structured components with at least one dimension less than 100nm. Materials that have one dimension in the nanoscale are layers, such as a thin films or surface coatings. Some of the features on computer chips come in this category. Materials that are nanoscale in two dimensions include nanowires and nanotubes. Materials that are nanoscale in three dimensions are particles, for example precipitates, colloids and quantum dots (tiny particles of semiconductor materials). Nanocrystalline materials, made up of nanometre-sized grains, also fall into this category

Nanomedecine

Application of nanoscience and nanotechnologies techniques in the field of medecine. Areas such as disease diagnosis, drug delivery and molecular imaging are being intensively researched. Medical-related products containing nanoparticles are currently produced

Nanometrology

Nanometrology is the science of measurement at the nanoscale level. Nanometrology has a crucial role in order to produce nanomaterials and devices with a high degree of accuracy and reliability.

Nanowires

A wire that connects transistors and gates that is less than 100 nanometers in size. They are ultrafine wires or linear arrays of dots, made from a wide range of materials.

Nanopowders

Nano-sized particles exhibit a range of physical, chemical and biological properties that are quite different to bulk materials of the same substance. Industry is now making use of these changed properties to enhance the functionality of many products

Nanophotonics

The manipulation and emission of light (both far-field and near-field) using nano-scale materials

Nanotransistor

A transistor that is less than 100 nanometers in size.

Nanotubes

A sheet of atoms arranged in the form of a tube.

Nanosurgery

A generic term including molecular repair and cell surgery

Optical Tunneling

A quantum mechanical phenomena resulting from photon delocalization that allows light to cross propagation barriers such as an interface.

Photolithography

Writing or creating patterns by means of light.

Typically, a photosensitive surface (a photoresist) is selectively exposed to light using a template. The exposed areas are subsequently etched (carved by chemical means). The technique used to produce the silicon chips that make up modern-day computers (actually, the term predates printed circuits, although that is the form we are interested in). The traditional process involves shining light through a mask onto a photosensitive polymer (photoresist) on a silicon surface, then subsequently removing the exposed areas.

Proximal probes

A family of devices capable of fine positional control and sensing, including scanning tunneling and atomic force microscopes; an enabling technology for nanotechnology

Quantum Dot

An object so small that adding or removing a single electron represents a significant change

Quantum Interferometric Lithography

A method of lithography that exploits quantum entanglement in order to focus a lithographic beam beyond half of its wavelength.

Quantum Mechanics

A physical model of chemical and optical phenomena, as well as the behaviour of matter in general on a small scale.

Quantum Mirage

A nanoscale property that may allow information to be transfered through use of the wave property of electrons.

Qubit

The quantum computing analog to a bit

Scanning Capacitance Microscopy (SCM)

A variation of scanning force microscopy that maps the localized capacitance of a surface.

Scanning Electron Microscopy (SEM)

A method of producing images of a surface by scanning an electron beam over the sample and measuring the electronic interactions with the interface.

Scanning Near-field Optical Microscopy (SNOM)

A type of scanning probe microscopy that maps the near-field optical properties of an interface.

Scanning Probe Microscopy (SPM)

A method for imaging nanoscale features of surfaces by scanning a sensor (probe) over a surface. Near-field effects such as tunneling, van der Waals forces, local fields and more are serially detected at localized points on the surface and used to create an SPM image.

Near-field effects such as tunneling, van der Waals forces, local fields and more are serially detected at localized points on the surface and used to create an SPM image

Scanning Thermal Microscopy

A type of scanning probe microscopy that maps the local temperature and thermal conductivity of an interface.

The probe in a scanning thermal microscope is sensitive to local temperatures—providing a nanoscale thermometer

Self-Assembled Monolayer (SAM)

A two-dimensional film, one molecule thick, covalently assembled at an interface.

The classical example of a SAM is the reaction of alkanethiols with a gold surface.

Self-Assembler

A specific type of assembler that makes use of self-assembly such that the assembly process would theoretically not require external energy or information input.

In a self-assembler, the choice of starting materials (input) determines the assembly and is always of higher energy than the product (output).

Scanning Tunneling Microscopy (STM)

A type of scanning probe microscopy that maps the local electrical properties of an interface.

Soft Lithography

A method for transferring a surface pattern that makes use of an elastomeric stamp.

Types of soft lithography include: Microcontact Printing (μCP), replica molding (REM), microtransfer molding (μTM), micromolding in capillaries (MIMIC) and solvent-assisted micromolding (SAMIM)

Spin

A quantum number describing the angular momentum and magnetic moment of a subatomic particle.

Spin-Spin Coupling

Interaction between the spin states of two subatomic particles.

Smart Materials

Materials and products capable of relatively complex behavior due to the incorporation of nanocomputers and nanomachines. Also used for products having some ability to respond to the environment.Programmable smart materials could shape-shift into just about any desired object. A house made of smart materials would be quite useful and interesting. Imagine a wall changing color at your command, or making a window where their was none before

Spintronics

Quantum Spintronics, Magnetoelectronics, Spin Electronics. Electronic devices that exploit the spin of electrons as well as their charge. Unlike conventional electronics which is based on number of charges and their energy, and whose performance limited in speed and dissipation, spintronics is based on the direction of electron spin, and spin coupling, and is capable of much higher speed at much lower power.

SNT

An abbreviation for structural nanotechnology; refers to integration of nanotech features into non-MNT products, also called nanomaterials

Scanning tunneling microscope (STM)

An instrument able to image conducting surfaces to atomic accuracy; has been used to pin molecules to a surface

Semiconductor

A semiconductor is a substance, usually a solid chemical element or compound, that has a conductivity between a metal and an isolator (e.g. Silicon, Germanium). It's conductivity is adjustable (doping) making it a good medium for the control of the electrical current.

Thermoelectric Material

A material capable of converting heat to electrical energy and electrical energy to heat.

Template

An external source of information or structure often used in manufacturing and mass production. As a nanoscale example, biological systems are known to use nucleic acids as a templates for protein synthesis.

Templating

The directed assembly of materials through interaction with a template.

Transmission Electron Microscopy (TEM)

A method of producing images of a sample by illuminating the sample with electronic radiation (in a vacuum), and detecting the electrons that are transmitted through the sample.

Blue Goo-opposite of Grey goo. Benificial tech, or "police" nanobots

Tissue engineering

The application of the principles and methods of engineering and the life sciences toward the fundamental understanding of structure/function relationships in normal and pathological mammalian tissues and the development of biological substitutes to restore, maintain, or improve functions.

Vasculoid

The vasculoid [concept] is a single, complex, multisegmented nanotechnological medical robotic system capable of duplicating all essential thermal and biochemical transport functions of the blood, including circulation of respiratory gases, glucose, hormones, cytokines, waste products, and cellular components

Wet Nanotechnology

The study of biological systems that exist primarily in a water environment. The functional nanometer-scale structures of interest here are genetic material, membranes, enzymes and other cellular components. The success of this nanotechnology is amply demonstrated by the existence of living organisms whose form, function, and evolution are governed by the interactions of nanometer-scale structures.

INDEX